The Impact of Environmental Variability on Ecological Systems

THE PETER YODZIS FUNDAMENTAL ECOLOGY SERIES

VOLUME 2

Series Editor

K. S. McCANN

The Impact of Environmental Variability on Ecological Systems

Edited by

D. A. VASSEUR
University of Calgary, AB, Canada

K. S. McCANN
University of Guelph, ON, Canada

Springer

A C.I.P. Catalogue record for this book is available from the Library of Congress.

ISBN-13 978-90-481-7457-7
ISBN-13 978-1-4020-5851-6 (e-book)

Published by Springer,
P.O. Box 17, 3300 AA Dordrecht, The Netherlands.

www.springer.com

Printed on acid-free paper

TABLE OF CONTENTS

List of Contributors

Peter A. Abrams
Department of Ecology and Evolutionary Biology
University of Toronto
25 Harbord St.
Toronto, Ontario M5S 3G5, Canada
abrams@zoo.utoronto.ca

Oscar De Feo
Department of Microelectronic Engineering
University College Cork
Butler Building
North Mall, Cork, Ireland

Mike S. Fowler
Integrative Ecology Unit
Department of Biological and Environmental Sciences
P.O. Box 65 (Viikinkaari 1), FIN-00014
University of Helsinki, Finland

Andrew Gonzalez
Department of Biology
McGill University
1205 ave Docteur Penfield
Montreal, QC H3A 1B1, Canada
andrew.gonzalez@mcgill.ca

John M. Halley
Department of Ecology
School of Biology
Aristotle University of Thessaloniki
U.P.B. 119,
54124 Thessaloniki, Greece
jmax@bio.auth.gr

Murray M. Humphries
Department of Natural Resource Sciences
Macdonald Campus
McGill University,
Ste-Anne-de-Bellevue
Quebec H9X 3V9, Canada
murray.humphries@mcgill.ca

Anthony R. Ives
Department of Zoology
UW-Madison
Madison, WI 53706, USA

Veijo Kaitala
Integrative Ecology Unit
Department of Biological and Environmental Sciences
P.O. Box 65 (Viikinkaari 1), FIN-00014
University of Helsinki, Finland

Jan Lindström
Graham Kerr Building
Division of Environmental and Evolutionary Biology
Institute of Biomedical and Life Sciences
University of Glasgow, Glasgow
G12 8QQ, Scotland

Valerie Loeb
Moss Landing Marine Laboratories
8272 Moss Landing Road
Moss Landing, California
USA, 95039
loeb@mlml.calstate.edu

Michael J. McPhaden
NOAA/Pacific Marine Environmental Laboratory
7600 Sand Point Way NE
Seattle, WA
98115 USA
Michael.j.mcphaden@noaa.gov

Owen L. Petchey
Department of Animal and Plant Sciences
University of Sheffield
Sheffield, S10 2TN
O.Petchey@Sheffield.ac.uk

Esa Ranta
Integrative Ecology Unit
Department of Biological and Environmental Sciences
P.O. Box 65 (Viikinkaari 1), FIN-00014
University of Helsinki, Finland
esa.ranta@helsinki.fi

Jörgen Ripa
Department of Theoretical Ecology
Ecology Building, Lund University
SE-223 62 Lund, Sweden
jorgen.ripa@teorekol.lu.se

James Umbanhowar
Department of Zoology
University of Guelph
Guelph Ontario
N1G 2W1, Canada

David A. Vasseur
Department of Biological Sciences
University of Calgary
Calgary, Alberta
Canada T2N 1N4
dvasseur@ucalgary.ca

Preface

This book, and the series in which it is published, arose from a series of ongoing colloquia aptly titled “Fundamental Ecology” and held in honour of the contributions of Peter Yodzis to the field of ecology. *The impact of environmental variability on ecological systems* was a topic selected by Peter for the 2005 colloquium. Peter succumbed to ALS (Amyotrophic Lateral Sclerosis) one month prior to the conference.

Introduction

Fluctuations in environmental conditions are ubiquitous across the globe; within terrestrial and aquatic environments, polar and equatorial environments, and alpine and ocean-deep environments, the physical conditions that influence life are continuously changing in time and space. The action of these fluctuations on an individual of any species can induce fluctuations in physiology, behaviour, phenology, and morphology (to name only a few), but it is their ultimate impact upon traits at the level of populations, communities, and ecosystems which ecologists most often seek to describe. Establishing links between the environment, the proximate, and the ultimate impacts of changing conditions will undoubtedly better our ability to understand and predict the outcome of natural and anthropogenic environmental changes.

One of the largest challenges for ecologists, when faced with the goal of understanding the impacts of environmental variability, is to determine which environmental characteristics to measure or model. While there are a few perhaps globally relevant environmental characters such as temperature, different habitats are limited to varying extents by different environmental characters. Across much of the planet primary production is influenced by solar irradiance, but there are indeed ecosystems existing deep in the oceans and in caves where light does not permeate; primary production in terrestrial systems can be limited by nitrogen and water availability, while aquatic systems primary production may be predominantly influenced by the availability of phosphorus, iron, and other "rare" elements. Heterotrophic organisms too experience environmental fluctuations which can directly affect their ability to acquire organic and inorganic resources: reductions in stream flow may limit the amount of suitable habitat for a certain fish species; wave action may simply wash away the residents of marine intertidal zones; and drought may desiccate even the sturdiest of terrestrial organisms. In some cases the indirect effects of changing environmental characteristics may be more important than the direct effects and in most cases their impact certainly cannot be neglected. When one considers how an environmentally driven change in primary production would impact the herbivores in an ecosystem by altering the availability of food or how a change in environmental conditions may facilitate or hamper the transmission of disease, it becomes immediately clear that a myriad of environmental characteristics may influence a single ecosystem.

Fluctuations in environmental characters occur upon two axes which are of equal and utmost importance for ecological processes, space, and time. At the regional level, environmental indices such as the North Atlantic Oscillation (NAO) and El

Niño Southern Oscillation (ENSO) have proven useful for explaining and predicting both the temporal and spatial variability in air and water temperature, precipitation, and storm frequency among many other characters (see McPhaden, Chapter 1, this volume ~ El Niño). As their names imply, these "oscillations" occur at semi-regular intervals in time, a typical feature of many of the forces driving fluctuations in the environment at the ecological timescale (e.g. seasonal, tidal, and diurnal cycles). In addition to the regularity of some such fluctuations, there is another feature of environmental variation that has proven widely important for ecology: the relatedness of proximate samples of a character across space and time. If spatially proximate sample measurements of an environmental character are highly related we expect the environment to be more homogeneous at smaller scales than it is at larger scales. Similarly in time, if proximate samples are highly related we expect the environment to vary less at shorter timescales. The commonality of these two characteristics of environmental variability, the periodicity, and relatedness of proximate samples, to many different environmental variables, makes them valuable topics for research and their importance resonates throughout the chapters in this book.

Examples of how environmental variables correlate with biological and physiological processes permeate the literature, but there are relatively few glimpses of truly causal relationships at the ecological level. Empiricists have, with recently increased frequency, turned to multivariate statistical methods such as principal components analysis (PCA) to determine the relatedness of changes in ecological and environmental characters. While these approaches have offered a variety of links between ecological and environmental processes, they lack the ability to answer the more fundamental question about how the environmental change is itself translated into the units in which we quantitatively measure ecological characters. Plaguing this basic understanding is the complexity with which any single environmental character can impact the ecology of a population or community. For example, in the short term individuals may express behavioural responses to changing environmental conditions, in the medium term changes in phenology (the timing of life-history events), and in the long term, evolutionary changes in their physiological and morphological machinery, all of these being potentially complex nonlinear functions. Add to this the interactions individuals have with con- and interspecifics, and the link between environmental and ecological fluctuations might be only faintly distinguishable. Yet, Ecology is making progress unmasking how different processes, independently and in concert, modulate the environmental signal into variability in ecological characters. This 2nd volume of Fundamental Ecology is devoted to highlighting recent progress along this avenue.

This book's predilection toward highly theoretical approaches comes about as a consequence of Peter Yodzis' influence upon the topic and participants of the 2005 colloquium in Fundamental Ecology and upon those of us who spent time as graduate students studying under Peter. The book begins with a contribution from Michael McPhaden highlighting the physical mechanisms responsible for, and the environmental consequences of the ENSO. This oscillation has what may arguably be the largest impact of any known recurrent environmental fluctuation outside of seasonal oscillations (at ecological timescales) and its impacts have been well-documented in a variety of ecological systems. Included in this is the Antarctic marine ecosystem which is detailed by Valerie Loeb in the final chapter of this book. Here the success of zooplankton and the entire ecosystem depends highly upon the extent of sea ice which is in turn related to the ENSO. Bridging these two chapters, which respectively describe the causes and consequences of environmental variability, are a diverse set of studies linked by their common goal of describing how different processes determine the impact of environmental variability on ecological systems.

The book continues with two studies which take an "energetic" approach to the problem of integrating the environment into ecological models. Murray Humphries and James Umbanhowar model the energy requirements of endotherms in fluctuating thermal environments and demonstrate that activity and the utilization of thermal refuges mediates their ultimate response to environmental variability. David Vasseur takes a similar approach to investigate how different frequencies of thermal fluctuations and body sizes of individuals influence the variability of communities. The importance of the temporal scale of fluctuations, which resonates throughout this book, is enhanced by a contribution from John Halley which demonstrates how our perceptions of population variability are influenced by the scale and resolution of the sampling regime. Esa Ranta et al. add to this discussion of temporal scale and variability, the impact of spatial scale upon the integration and expression of variability in ecological systems, highlighting specifically the prevalence and impact of population synchrony that has been found in models and empirical studies.

The remaining chapters investigate how interactions between populations impact their uptake and expression of fluctuations and upon processes at higher levels of organization. Jörgen Ripa and Anthony Ives demonstrate that the relationship between populations is easily obscured in fluctuating environments and they develop a framework for identifying these interactions. One such interaction, namely competition, is highlighted by Peter Abrams as a process with the potential to influence patterns of species coexistence in fluctuating environments. Contributions from Andrew Gonzalez and Oscar De Feo, and from Owen Petchey, respectively use theoretical models and microcosm experiments to investigate how diversity may provide "insurance" against the detriments of existing in a fluctuating environment.

Environmental variability is such a ubiquitous aspect of ecology that we cannot hope to provide a comprehensive overview of the current related research. Rather we have tried in earnest to provide studies from a range of related niches which emphasize the importance of environmental and ecological processes spanning from the generation of environmental fluctuations, their uptake by ecological processes, and their expression in ecological characteristics. In the coming years, anthropogenic impacts will undoubtedly continue to alter the expression of environmental variability, challenging our understanding of ecological processes and our ability to restore, manage, and defend ecological systems.

CHAPTER 1

EL NIÑO AND LA NIÑA: PHYSICAL MECHANISMS AND CLIMATE IMPACTS

MICHAEL J. MCPHADEN

NOAA/Pacific Marine Environmental Laboratory, 7600 Sand Point Way NE, Seattle, WA 98115 USA
Phone: (206) 526-6783, Fax: (206) 526-6815, E-mail: Michael.j.mcphaden@noaa.gov

1.1 ABSTRACT

This paper summarizes the physical mechanisms responsible for year-to-year variations associated with the El Niño–Southern Oscillation (ENSO) cycle of warm (El Niño) and cold (La Niña) events originating in the tropical Pacific. The global climatic impacts of ENSO are also reviewed. The paper concludes with a discussion of outstanding research issues related to the dynamics of ENSO, its interactions with other modes of natural climate variability, and its possible modification by global warming.

Keywords: El Niño–Southern Oscillation, Climate Impacts, weather, ecosystem.

David A. Vasseur and Kevin S. McCann, The Impact of Environmental Variability on Ecological Systems, 1–16.

1.2 INTRODUCTION

The El Niño–Southern Oscillation (ENSO) cycle is the most prominent year-to-year climate fluctuation on Earth. It originates in the tropical Pacific with unusually warm (El Niño) and cold (La Niña) events recurring approximately every 2–7 years. El Niño and La Niña events typically last 12–18 months and develop in association with swings of atmospheric pressure between the tropical Indo-Pacific and eastern Pacific. These pressure swings, known as the Southern Oscillation, are intimately related to the strength of the Pacific trade winds. ENSO extends its reach beyond the tropical Pacific through atmospheric teleconnections that affect patterns of weather variability worldwide. As one indicator of socioeconomic impact, the 1997–1998 El Niño, by some measures the strongest of the 20th century (McPhaden 1999), resulted in 22,000 fatalities and US$36 billion in economic losses worldwide (Sponberg 1999).

The altered environmental conditions that result from El Niño and La Niña influence global patterns of primary production (the fixation of carbon by plants) (Behrenfeld 2001), with effects that ripple through higher levels of the food chain in both marine and terrestrial ecosystems (Stenseth et al. 2002). Oceanic consequences include changes in the mortality, fecundity, and/or geographic distribution of marine mammals, sea birds, and commercially valuable fish species (Glantz 2001). In addition, elevated sea temperatures may contribute to coral bleaching (Glynn 1988; Lough 2000). On land, seasonal shifts in the pattern of rainfall, air temperature, and sunlight availability during El Niño and La Niña can have dramatic effects on plants and animals in such diverse environments as tropical rainforests, mangrove swamps, boreal forests, deserts, and semi-arid shrub lands (Holmgren et al. 2001, 2006).

The purpose of this paper is to provide a brief overview of the physics governing El Niño and La Niña events and how these events affect global climate. By climate, we mean statistical averages or other statistical properties of variations in the ocean-atmosphere-land system over periods of 1 month or more. Climate thus involves slowly varying aspects of the Earth system as opposed to, for example, the details of individual weather events that occur on timescales of minutes to days.

The reader is referred to McPhaden et al (2006) for a review of ENSO that cuts across a broad range of topics in Earth and related Sciences. Special focus reviews of ENSO impacts on ecosystems can be found in Glynn (1988), Stenseth et al. (2003), and Holmgren et al (2006). Kovats et al. (2004) review the impacts of ENSO variability on human health while Glantz (2001) provides historical and socioeconomic perspectives on the ENSO phenomenon. Goddard et al. (2001) summarize the status of ENSO forecasting efforts, which are motivated by the desire to mitigate the adverse impacts or to take advantage of the altered environmental conditions associated with El Niño and La Niña events.

1.3 ENSO PHYSICS

Jacob Bjerkes, a Norwegian meteorologist, was the first to recognize that El Niño results from coupled interactions between the ocean and the atmosphere, mediated by trade wind and sea surface temperature (SST) variations in the equatorial Pacific (Bjerknes 1966, 1969). He was also the first to plausibly link year-to-year variations in the Southern Oscillation to El Niño. These seminal insights are the foundation for much of the research on ENSO dynamics over the last 40 years.

To understand how El Niño works, we must first describe how the tropical Pacific ocean-atmosphere system behaves under normal conditions. Typically, the trade winds along the equator drive surface flow westward in the South Equatorial Current (SEC) (Figure 1.1). This current drains surface water heated by the intense tropical sun from the eastern Pacific and piles it up in the western Pacific. The thermocline, which is the region of sharp vertical temperature gradient separating the warm surface layer from the cold deep ocean, is pushed down in the west and elevated in the east. Sea level tends to mirror thermocline depth since sea water expands when heated. Thus, while the thermocline tilts downward toward the west by 100 m along the equator, sea level stands about 60 cm higher in the western Pacific than in the eastern Pacific.

The relative shallowness of the thermocline in the eastern Pacific facilitates the upwelling (i.e. upward transport) of cold interior water by the trade winds, and a cold tongue develops in SST from the coast of South America to near the International Date Line. The east–west surface temperature contrast reinforces the easterly trade winds which near the equator flow from a region of high surface air pressure in hydrostatic equilibrium with underlying cold water in the east to a region of low surface air pressure over warm water in the west. This positive feedback between trade wind intensity and zonal SST contrasts is referred to as the Bjerknes feedback (Cane 2005).

As the trade winds flow from east to west, they pick up heat and moisture from the ocean. The warm, humid air mass becomes less dense and rises over the western Pacific warm pool (SST $\geq$ 28°C) where deep convection leads to towering cumulus clouds and heavy precipitation. Ascending air masses in this region of deep convection return eastward in the upper levels of the troposphere, then sink over the cooler water of the eastern Pacific. This circulation loop is often referred to as the Walker Circulation in honor of Sir Gilbert Walker who first described the Southern Oscillation in the early 20th century (Bjerknes 1969).

During El Niño, the trade winds weaken along the equator as atmospheric pressure rises in the western Pacific and falls in the eastern Pacific. Weaker trade winds allow the western Pacific warm pool to migrate eastward and the thermocline to flatten out across the basin (Figure 1.1). The ability of upwelling to cool the surface is reduced where the thermocline deepens. Hence, surface temperatures warm significantly in the eastern and central equatorial Pacific because of the combination of reduced cooling by upwelling and eastward shifts in the warm pool.

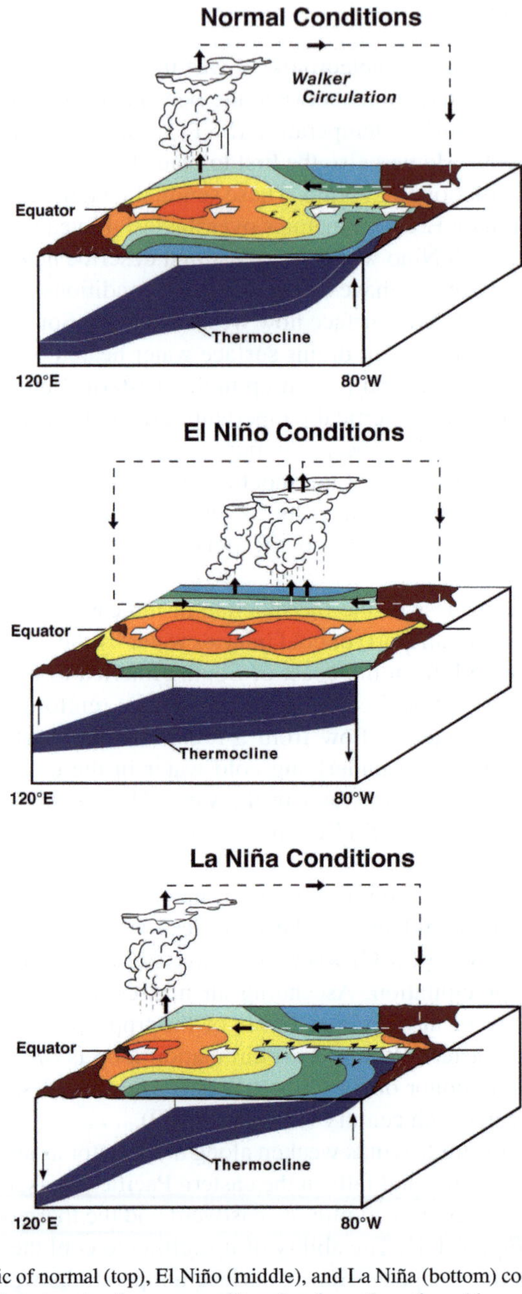

Figure 1.1. Schematic of normal (top), El Niño (middle), and La Niña (bottom) conditions in the tropical Pacific. The thermocline separates the warm sunlit surface layer from the cold ocean interior. The Walker Circulation is the atmospheric circulation on the equatorial plane that involves rising air masses and rainfall over warm water and sinking air masses over cold water.

In the atmosphere, deep cumulus clouds and heavy rains associated with the ascending branch of the Walker Circulation normally occur in the western tropical Pacific over the warmest water. As surface temperatures warm to the east during El Niño, convective cloudiness and rainfall migrate eastward toward the date line. This eastward shift in deep convection favors further weakening of the trade winds, since in the western Pacific surface winds flowing into the region of convection are from the west (Figure 1.1). Thus, the Bjerknes feedback that operates under normal conditions now runs in reverse, with weakened trade winds and SST warming tendencies along the equator reinforcing one another as El Niño develops (Rasmusson and Wallace 1983).

The flip side of El Niño, known as La Niña, is characterized by stronger-than-normal trade winds and colder-than-normal tropical Pacific SSTs (Figure 1.1). Because of the stronger trade wind forcing, the thermocline slopes down to the west more steeply than usual along the equator. Also, the zonal extent of the western Pacific warm pool is reduced and heavy rainfall is confined to the far western tropical Pacific because of the westward penetration of colder-than-normal SSTs along the equator (Figure 1.1).

The ENSO theory emphasizes the importance of oceanic waves generated by large-scale variations in surface winds for understanding the cycle of warm and cold events in the tropical Pacific (Neelin et al. 1998; Wang and Picaut 2004). The Earth's rotational forces trap these waves within several hundred kilometers of the equator in the open ocean, so they transfer energy very efficiently over many thousands of kilometers in the east–west direction. Equatorial waves are responsible for flattening the thermocline in response to weakened trade winds during the onset of El Niño and for steepening the thermocline in response to strengthened trade winds during the onset of La Niña (Figure 1.1).

As El Niño events mature, currents in the meridional (i.e. north–south) direction associated with these waves drain warm surface layer water out of the equatorial band and transport it to higher latitudes. Loss of warm water causes the equatorial thermocline to slowly shoal across the basin, creating conditions favorable for SST to start cooling in the east. This cooling signals the end of El Niño and, if strong enough, the onset of La Niña (Jin 1997; Meinen and McPhaden 2000). Similarly, prior to the onset of El Niño, meridional currents pump warm surface layer water into the equatorial band from neighboring latitudes, pushing the thermocline down across the basin to create conditions favorable for subsequent warming in the cold tongue. The time it takes for these wave-related meridional currents to slowly transport warm water into and out of the equatorial band is important in determining the interval between ENSO warm and cold phases.

SST patterns (Figure 1.2) illustrate equatorial warm pool and eastern Pacific cold tongue structures for three different Decembers during the boreal winter season when El Niño and La Niña events typically reach their peak development. Anomalies (i.e. deviations from normal) highlight the characteristic warming of the cold-tongue region during El Niño and enhanced cooling during La Niña (Figure 1.3). The occurrence of warm and cold ENSO events over the last 55 years is shown in the plot of

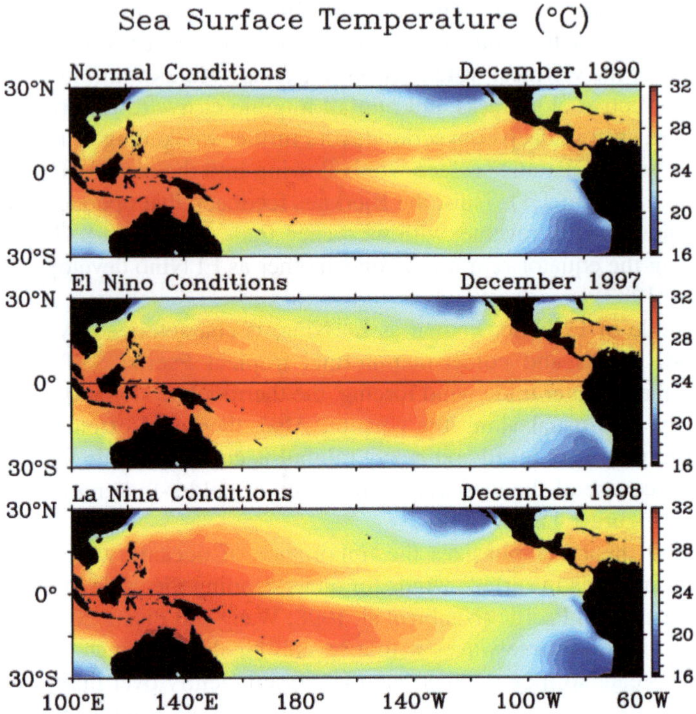

Figure 1.2. Sea surface temperatures for December 1990 (a normal year), December 1997 (an El Niño year), and December 1998 (a La Niña year).

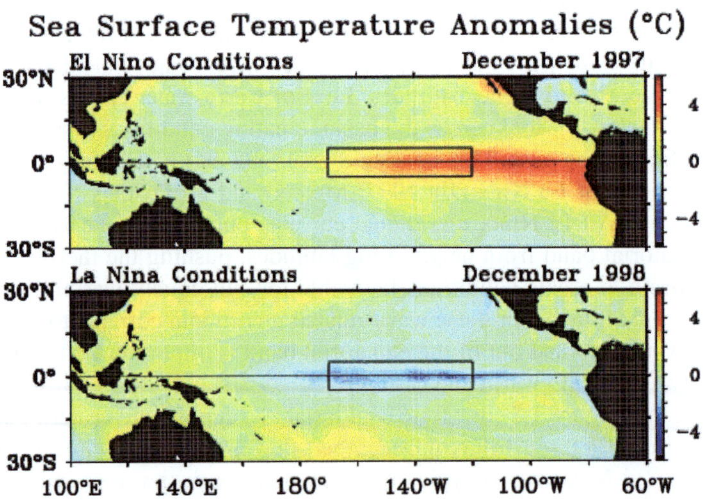

Figure 1.3. Sea surface temperature anomalies for December 1997 and December 1998. The boundaries of the Niño-3.4 index region are superimposed.

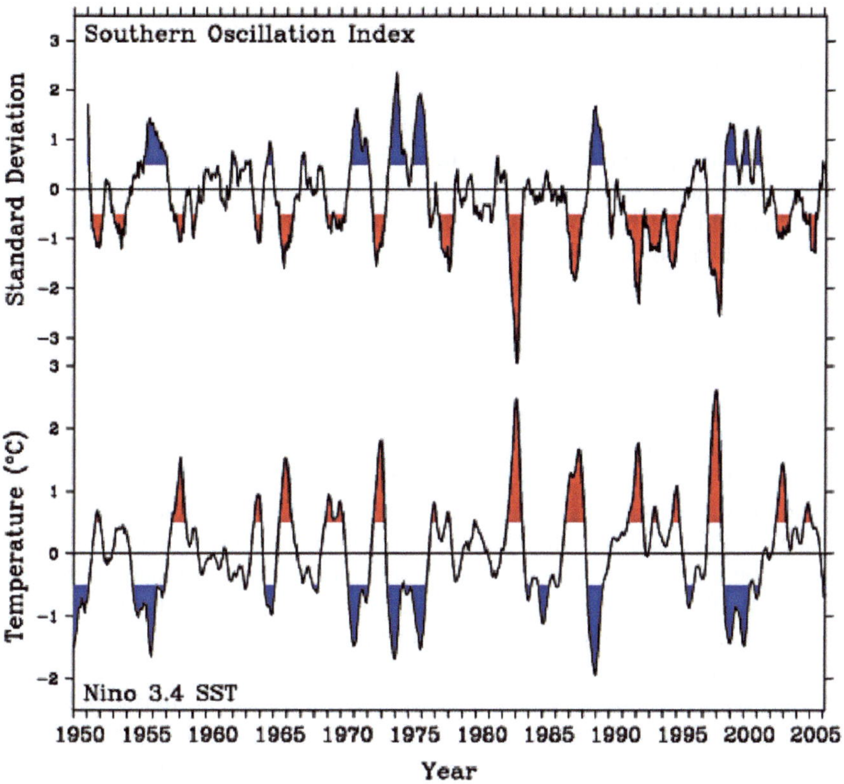

Figure 1.4. The Southern Oscillation index (SOI) and Niño-3.4 sea surface temperature index for 1950 to 2005. The Niño-3.4 index is computed from monthly SST anomalies in the region 5°N–5°S, 120°–170°W. Positive anomalies greater than 0.5°C indicate El Niño events, and negative anomalies less than −0.5°C indicate La Niña events. The SOI represents the intensity of the easterly trade winds and is computed as a normalized surface air pressure difference between Tahiti, French Polynesia, minus Darwin, Australia, after mean seasonal variations have been filtered out. Periods of SOI greater in magnitude than 0.5 are shaded to emphasize the relationship with El Niño and La Niña. Low SOI is associated with weaker trade winds and warm sea temperatures (El Niño) while high SOI is associated with stronger trade winds and cold sea temperatures (La Niña). All values have been smoothed with a 5-month running mean for clarity.

SST anomalies from the Niño-3.4 region for the period 1950–2005 (Figure 1.4). The strongest El Niños of the last 55 years occurred in 1982–1983 and 1997–1998, while recent strong La Niñas occurred in 1988–1989 and 1998–2000. The Southern Oscillation Index (SOI), defined as the normalized difference in surface atmospheric pressure between Tahiti, French Polynesia, and Darwin, Australia, is a measure of trade wind strength and shows nearly perfect anticorrelation with the Niño-3.4 index (Figure 1.4). The tight relationship between SOI and Niño-3.4 SST emphasizes the significant coupling between the ocean and the atmosphere that gives rise to ENSO variations.

1.4 CLIMATE IMPACTS

One impact of the eastward shift in rainfall along the equator during El Niño is that drought typically develops in Australia, Indonesia, and the neighboring countries. On the other hand, the island states of the central Pacific and the west coast of South America are often inundated with heavy rains. Heavy rainfall bands normally situated north and south of the equator in the Intertropical Convergence Zone and the South Pacific Convergence Zone also shift equatorward as surface waters warm. These latitudinal shifts contribute to unusually heavy rains near the equator in the central and eastern Pacific, and to drought conditions at higher latitudes in regions such as New Caledonia and Fiji to the south, and Hawaii to the north. Droughts also tend to occur in Northeast Brazil, the Amazon, and South Africa. Likewise, Indian summer monsoon rainfall typically decreases during an El Niño year.

Heat released into the troposphere from moisture condensation in regions of deep tropical convection is one of the principal driving forces for the global atmospheric circulation. Changes in the location and intensity of these heat sources during El Niño therefore lead to widespread changes in wind and weather patterns outside the tropical Pacific. The remote effects of El Niño are referred to as teleconnections. In the extratropics, teleconnections are usually most pronounced during the late boreal fall and winter season when ENSO SST anomalies tend to be largest. However, notable teleconnections, some of which are discussed below, are detectable in other seasons as well.

Anomalous tropospheric heating in the central tropical Pacific during El Niño generates quasi-stationary atmospheric wave trains that radiate poleward and eastward. In the northern hemisphere, these waves set up the Pacific North American (PNA) teleconnection pattern, which is a series of anomalous high- and low-pressure centers extending from the central North Pacific to North America (Figure 1.5). During El Niño, the Aleutian low pressure center over the North Pacific deepens, high pressure develops over western North America, and low pressure prevails over the southeastern USA. These pressure changes steer warmer air masses from southern latitudes into the Pacific Northwest and southern Canada, so that El Niño winters tend to be mild over much of Canada and the northwestern USA. Low pressure in the southeastern USA associated with the PNA pattern brings wetter, cooler conditions to the states bordering the Gulf of Mexico. Similar wave trains are excited in the southern hemisphere, but they are weaker and more variable than those in the north (Trenberth et al. 1998).

Teleconnections also affect the subtropical jet streams, which are swift air flows girdling the Earth at altitudes centered between about 10,000 and 12,000 m. The eastward migration of deep convection during El Niño causes the core of these jet streams to intensify and shift equatorward and eastward in the central and eastern Pacific (Figure 1.5). The strong winds of the jet streams steer storms that form near their axes like leaves in a stream, creating identifiable storm tracks that mark regions of enhanced weather variability in the atmosphere. Hence, as a result of the equatorward shift in the jet streams during El Niño, southern California and northern Chile

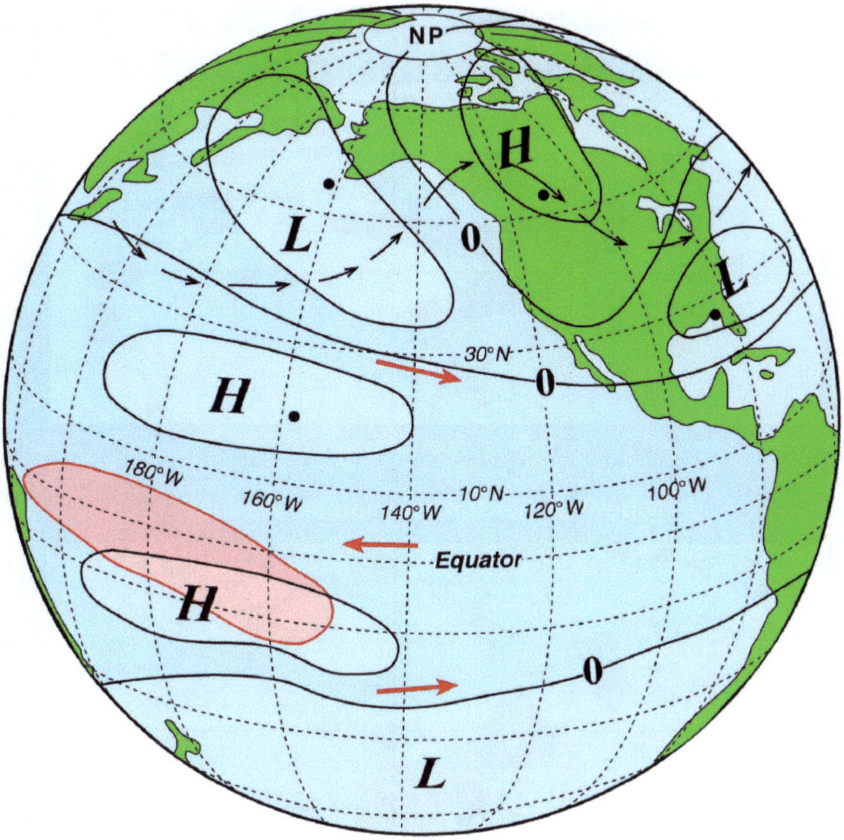

Figure 1.5. Schematic diagram of the Pacific North American (PNA) pattern of middle tropospheric pressure anomalies during the northern hemisphere winter coinciding with an El Niño event. Eastward pointing red arrows indicate strengthening of the subtropical westerly jet streams. The westward pointing red arrow near the equator reflects the shift in the Walker circulation between normal and El Niño conditions shown in Figure 1.1. Light black arrows depict mid-tropospheric air flows as distorted by the anomalous high-and low-pressure patterns. Rainfall and cloudiness are enhanced over the shaded region near the equator. (After Horel and Wallace 1981.)

typically experience stormier and wetter weather in their respective winters when the jet stream is seasonally most intense.

Global composites of typical rainfall and air temperature anomalies associated with El Niño during the boreal winter (December–February) and summer (June–August) seasons (Figure 1.6) illustrate many of the anomalous shifts in weather patterns discussed in the preceding paragraphs. La Niña often produces effects on global patterns of weather variability that are roughly opposite to those of El Niño (Halpert and Ropelewski 1992) although asymmetries in convective heating of the tropical atmosphere favor more intense responses to strong El Niños than strong La Niñas

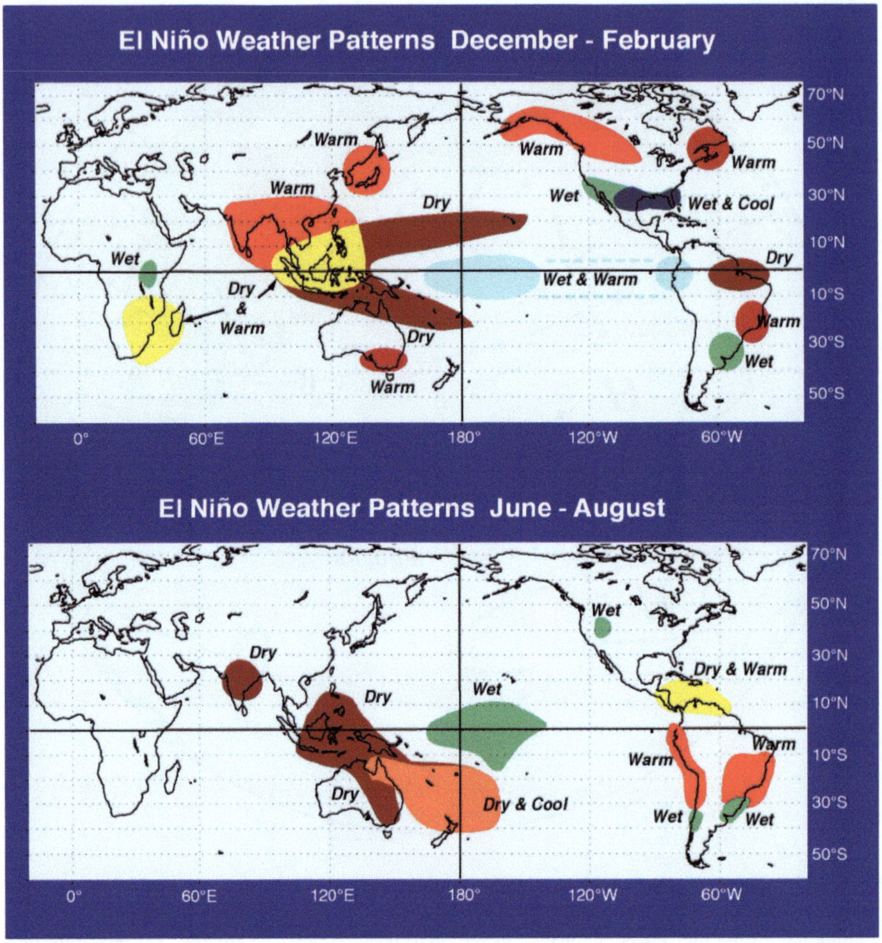

Figure 1.6. Schematic diagram showing temperature and precipitation anomalies associated with El Niño during the boreal winter and summer seasons. To a first approximation, the impacts of La Niña are similar, but with opposite sign. (After Ropelewski and Halpert 1987.)

(Hoerling et al. 2001). Not shown in these composites are variations in the climate of the Antarctic continent and surrounding seas, where atmospheric teleconnections emanating from the tropical Pacific lead to seasonal anomalies in air and sea temperatures, winds, atmospheric pressure, sea ice, and precipitation (Turner 2004; Yuan 2004). ENSO also has a detectable impact on European rainfall during the boreal spring season following peak tropical Pacific SST anomalies (van Oldenborgh et al. 2000).

Teleconnections affect surface conditions over land and the ocean that can feed back to the atmosphere to reinforce ENSO variations. At extratropical latitudes of

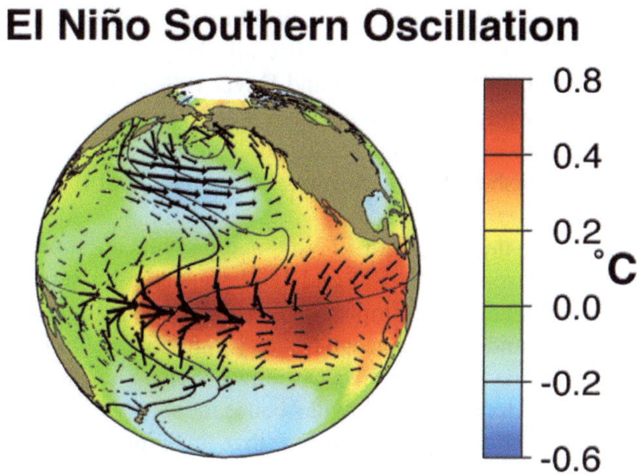

Figure 1.7. Anomalies of sea surface temperature (color shading), surface atmospheric pressure (contours: solid positive, dashed as negative), and surface winds (vectors) in the Pacific basin linked to El Niño events. Anomalies of the opposite sign apply to La Niña events. (After Mantua et al. 1997.) (Courtesy of S. Hare, University of Washington.)

the North Pacific, for example, SSTs cool in response to intensified surface westerlies (Figure 1.7), tending to strengthen the PNA pressure pattern (Lau and Nath 1996). Over land, changes in temperature and precipitation can affect soil moisture and evaporation, factors that contribute to the severity and duration of droughts and floods. Trenberth and Guillemot (1996) argued that these land surface processes helped to perpetuate La Niña-related drought in the US Midwest in 1988 and El Niño-related flooding in the US Midwest during 1993.

El Niño and La Niña also affect the frequency, intensity, and spatial distribution of tropical storms (Gray 1984). For example, during moderate to strong El Niño events, Atlantic hurricane formation tends to be suppressed because the intensified subtropical jet stream shears off the top of fledgling storms before they fully develop. In contrast, during La Niña, Atlantic hurricanes tend to be stronger and more numerous. These year-to-year changes in ENSO-related hurricane activity translate into a 3-to-1 greater likelihood of a major Atlantic hurricane making landfall in the USA during La Niña vs. El Niño years. Correspondingly, there is a higher probability of economic losses and threats to public safety during La Niña years (Pielke and Landsea 1999).

El Niño affects the coastal zones of the Americas all the way to Alaska in the northern hemisphere and to central Chile in the southern hemisphere through oceanic teleconnections (Figure 1.7). When eastward propagating oceanic waves generated by weakening of the trade winds at the onset of El Niño reach the coast of South America, part of the wave energy radiates northward and southward as coastal waves. These waves raise sea level and push the thermocline down, reducing the upwelling of cool water as they progress poleward. Within 1–2 months after the first equatorial waves reach South America, SSTs rise all along the coast to the north and south (Enfield and Allen 1980; Chelton and Davis 1982). North of central California, southerly surface winds related to the PNA pattern (Figure 1.7) also contribute to coastal warming (Ramp et al. 1997). These winds bring warm subtropical air northward and also drive onshore ocean currents that converge at the coast to push the thermocline down.

Geographically, climate impacts of El Niño and La Niña are most consistent from event to event in the tropical Pacific and bordering areas where the atmosphere responds directly to SST forcing. Impacts are prominent but sometimes less consistent at higher latitudes and in other ocean basins where regional influences or chaotic weather fluctuations may obscure ENSO signals. Strong events like 1982–1983 and 1997–1998 have dramatic worldwide consequences while weak events as in 2004–2005 (Figure 1.4) may have impacts that are muted or even undetectable above the background noise of the atmosphere (Lyon and Barnston 2005). Thus, although El Niño and La Niña increases the probability of a particular kind of weather pattern occurring in various regions of the globe, actual impacts may vary from those expected for any given event (Trenberth et al. 1998).

1.5 DISCUSSION

Much has been learned over the last 30 years about ENSO dynamics and its predictability, but there are still many unanswered questions about the nature of El Niño, La Niña, and the ENSO cycle. There is considerable debate at present, for example, about what causes the trade winds to relax at the onset of El Niño. One perspective is that ENSO freely oscillates between preferred warm and cold phases as part of a continuum in which weakening of the trades during the onset of El Niño is the result of large-scale deterministic processes operating during previous phases. Another perspective is that the ocean and atmosphere in the tropical Pacific tend to stably reside in a preferred state of cold in the east and warm in the west and that El Niño events occur only when the system is energized by forcing in the form of episodic westerly wind bursts. There is considerable observational evidence that episodic wind forcing plays an important role in the development of El Niño events (McPhaden 1999, 2004; Vecchi and Harrison 2000; Eisenman et al. 2005) but whether this wind forcing is an essential triggering mechanism or simply adds irregularity to an otherwise regular cycle is unresolved (Moore and Kleeman 1999; Fedorov and Philander 2000).

It is also important to recognize the broader context of natural and human-induced impacts on ecosystems when interpreting and assessing the role of ENSO variability. There are, for instance, many other modes of natural climate variability that manifest themselves in seasonal and longer timescale environmental changes. To mention just a few, the North Atlantic Oscillation (NAO; Hurrell et al. 2001) and the Pacific Decadal Oscillation (PDO; Mantua et al. 1997) involve basin-scale fluctuations in SSTs, air temperatures, atmospheric pressure, winds, and rainfall patterns. The recently discovered Indian Ocean dipole (Webster et al. 1999; Saji et al. 1999) represents an El Niño-like oscillation of the coupled ocean-atmosphere system in the tropical Indian Ocean while the interhemispheric SST gradient mode in the Atlantic links oceanic variability to drought over Northeast Brazil (Nobre and Shukla 1996) and the Sahel (Folland et al. 1986).

ENSO interacts with these and other modes of natural climate variability in complicated ways. For example, the PDO is sometimes referred to as "ENSO-like decadal variability" because of similarities with ENSO in terms of spatial patterns in surface winds, atmospheric pressure, and SST. The PDO was in a warm phase in the tropical Pacific from the mid-1970s to the late 1990s, which may have favored the extended period of relatively strong and long-lived El Niño events evident in Figure 1.4 (Fedorov and Philander 2000). However, alternative interpretations for this unusual El Niño behavior exist and whether the PDO is the cause of or consequence of the decadal modulation of ENSO remains an area of active research (Rodgers et al. 2004).

Interaction with or interference from other natural modes of climate variability can also affect ENSO impacts. For example, climate variations intrinsic to the Indian Ocean and the Atlantic Ocean compete with ENSO influences on Indian summer monsoon rainfall (Webster et al. 1999) and rainfall in Northeast Brazil (Enfield 1996). Also, ENSO-related rainfall variability over the USA (Gershunov and Barnett 1998) and Australia (Power et al. 1999) is markedly different during warm- and cold-phase PDO conditions.

Trenberth and Hoar (1996) cited the tendency for more frequent El Niños in the last 25 years of the 20th century and the extended period of unusual warmth from 1990–1995 (Figure 1.4) as evidence that global warming due to greenhouse gas emissions has affected the ENSO cycle. Careful examination of historical and paleoclimate records though indicates that this recent behavior is most likely not outside the range expected for natural climate variability (Wunsch 1999; Tudhope et al. 2001). Competing hypotheses, such as random fluctuations in atmospheric forcing or interaction with the PDO are equally plausible (Fedorov and Philander 2000).

Whether or not *future* global warming will affect ENSO is open to debate. It has been suggested that the frequency and amplitude of ENSO variability may increase (Timmermann et al. 1999) or that the tropical Pacific may develop either permanent El Niño-like (Meehl and Washington 1996) or La Niña-like (Cane et al. 1997) conditions in a warmer world. However, the consensus from current generation of global coupled ocean-atmosphere climate models is that there will be neither significant change in ENSO characteristics nor a shift towards ENSO-like background conditions in the tropical Pacific under various greenhouse gas emission scenarios that presume a doubling of atmospheric CO_2 from preindustrial levels over the next 100 years (van Oldenborgh et al. 2005). The models have known flaws however that affect their ability to accurately represent important details of climate variability and change. Thus, the possibility that global warming may affect ENSO in the future cannot be ruled out at present.

1.6 CONCLUSION

Our understanding of ENSO and its global climate impacts has significantly deepened and broadened in the last 20 years. This improved understanding provides a valuable framework for pursuing interdisciplinary studies of ENSO's influence on marine and terrestrial ecosystems. These studies are especially relevant today in view of international efforts to develop informed policies for sustainable development and responsible stewardship of the environment.

1.7 ACKNOWLEDGMENTS

The author would like to acknowledge Milena Holmgren for very helpful comments on an earlier version of this manuscript.

1.8 LITERATURE CITED

Behrenfeld, M.J. et al. 2001. Biospheric primary productivity during and ENSO transition. *Science* 291: 2594–2597.

Bjerknes, J. 1966. A possible response of the atmospheric Hadley circulation to equatorial anomalies of ocean temperature. *Tellus* 18: 820–829.

Bjerknes, J. 1969. Atmospheric teleconnections from the equatorial Pacific. *Monthly Weather Review* 97: 163–172.

Cane, M.A. et al. 1997. Twentieth century sea surface temperature trends. *Science* 275: 957–960.

Cane, M.A. 2005. The evolution of El Niño, past and future. *Earth and Planetary Science Letters* 230: 227–240.

Chelton, D.B. and R.E. Davis. 1982. Monthly mean sea-level variability along the west coast of North America. *Journal of Physical Oceanography* 12: 757–784.

Eisenman, I., L. Yu, and E. Tziperman. 2005. Westerly wind bursts: ENSO's tail rather than the dog? *Journal of Climate* 18: 5224–5238.

Enfield, D.B. 1996. Relationships of inter-Atlantic rainfall to tropical Atlantic and Pacific SST variability. *Geophysical Research Letters* 23: 3505–3508.

Enfield, D.B. and J.S. Allen. 1980. On the structure and dynamics of monthly mean sea level anomalies along the Pacific coast of North and South America. *Journal. of Physical Oceanography* 10: 557–578.

Fedorov, A.V. and S.G. Philander. 2000. Is El Niño changing? *Science* 288: 1997–2001.

Folland, C., T. Palmer, and D. Parker. 1986. Sahel rainfall and worldwide sea temperatures: 1901–1985. *Nature* 320: 602–606.

Gershunov, A. and T.P. Barnett. 1998. Interdecadal modulation of ENSO teleconnections. *Bulletin of the American Meteorological Society* 79: 2715–2725.

Glantz, M.H. 2001. *Currents of change: El Niño's impact on climate and society*, Cambridge University Press, Cambridge.

Glynn, P.W. 1988. El Niño-Southern Oscillation 1983–1983: nearshore population, community, and ecosystem responses. *Annual Review of Ecology and Systematics* 19: 309–345.

Goddard, L., S.J. Mason, S.E. Zebiak, C.F. Ropelewski, R. Basher, and M.A. Cane. 2001. Current approaches to seasonal to interannual climate predictions. *International Journal of Climatology* 21: 1111–1152.

Gray, W.M. 1984. Atlantic hurricane frequency: El Niño and 30mb quasi-biennial oscillation influences. *Monthly Weather Review* 112: 1649–1668.

Halpert, M.S. and C.F. Ropelewski. 1992. Surface temperature patterns associated with the Southern Oscillation. *Journal of Climate* 5: 577–593.

Hoerling, M.P., A. Kumar, and T. Xu. 2001. Robustness of the nonlinear climate response to ENSO's extreme phases. *Journal of Climate* 14: 1277–1293.

Holmgren, M., M. Scheffer, E. Ezcurra, J.R. Guiterrez, and G.M.J. Mohren. 2001. El Niño effects on the dynamics of terrestrial ecosystems. *Trends in Ecology Evolution* 16: 89–94.

Holmgren, M. et al. 2006. Extreme climate events shape arid and semi-arid ecosystems. *Frontiers in Ecology and the Environment* 4: 87–95.

Horel, J.D. and J.M. Wallace. 1981. Planetary scale atmospheric phenomena associated with the Southern Oscillation. *Monthly Weather Review* 109: 813–829.

Hurrell, J.W., J. Kushnir, and M. Visbeck. 2001. The North Atlantic Oscillation. *Science* 291: 603–605.

Jin, F.F. 1997. An equatorial recharge paradigm for ENSO. Part I: conceptual model. *Journal of Atmospheric Sciences* 54: 811–829.

Kovats, R.S., M.J. Bouma, S. Hajat, E. Worrall, and A. Haines. 2004. El Niño and health. *Lancet* 362: 1481–1489.

Lau, N.-C. and M.J. Nath. 1996. The role of the "atmospheric bridge" in linking tropical pacific ENSO events to extratropical SST anomalies. *Journal of Climate* 9: 2036–2057.

Lough, J.M. 2000. 1997–1998: unprecedented thermal stress to coral reefs? *Geophysical Research Letters* 27: 3901–3904.

Lyon, B. and A.G. Barnston. 2005. The evolution of the weak El Niño of 2004–2005. U.S. CLIVAR Variations. US CLIVAR Office, Washington, DC. **3**(2): 1–4.

Mantua, N.J., S.J. Hare, Y. Zhang, J.M. Wallace, and R.C. Francis. 1997. A Pacific interdecadal oscillation with impacts on salmon production. *Bulletin of the American Meteorological Society* 78: 1069–1079.

McPhaden, M.J. 1999. Genesis and evolution of the 1997–1998 El Niño. *Science* 283: 950–954.

McPhaden, M.J. 2004. Evolution of the 2002–2003 El Niño. *Bulletin of the American Meteorological Society* 85: 677–695.

McPhaden, M.J., S.E. Zebiak, and M.H. Glantz. 2006. ENSO as an integrating concept in Earth Science. *Science* 314: 1740–1745.

Meehl, G.A. and W.M. Washington. 1996. El Niño-like climate change in a model with increased atmospheric CO_2 concentrations. *Nature* 382: 56–60.

Meinen, C.S. and M.J. McPhaden. 2000. Observations of warm water volume changes in the equatorial Pacific and their relationship to El Niño and La Niña. *Journal of Climate* 13: 3551–3559.

Moore, A.M. and R. Kleeman. 1999. Stochastic forcing of ENSO by the intraseasonal oscillation. *Journal of Climate* 12: 1199–1220.

Neelin, J.D., D.S. Battisti, A.C. Hirst, F.-F. Jin, Y. Wakata, T. Yamagata, and S. Zebiak. 1998. ENSO theory. *Journal of Geophysical Research* 103: 14261–14290.

Nobre, P. and J. Shukla. 1996. Variations of sea surface temperature, wind stress and rainfall over the tropical Atlantic and South America. *Journal of Climate* 9: 2464–2479.

Pielke, R.A., Jr. and C.N. Landsea. 1999. La Niña, El Niño, and hurricane damages in the United States. *Bulletin of the American Meteorological Society* 80: 2027–2033.

Power, S., T. Casey, C. Folland, A. Colman, and V. Mehta. 1999. Inter-decadal modulation of the impact of ENSO on Australia. *Climate Dynamics* 15: 319–324.

Ramp, S.R., MJ.L. McClean, C.A. Collins, A.J. Semtner, and K.A.S. Hayes. 1997. Observations and modeling of the 1991–1992 El Niño signal off central California. *Journal of Geophysical Research* 102: 5553–5582.

Rasmusson, E.M. and J.M. Wallace. 1983. Meteorological aspects of the El Niño/Southern Oscillation. *Science* 222: 1195–1202.

Rodgers, K.B., P. Freiderichs, and M. Latif. 2004. Tropical Pacific decadal variability and its relation to the decadal modulation of ENSO. *Journal of Climate* 17: 3761–3774.

Ropelewski, C.F. and M. Halpert. 1987. Global and regional scale precipitation patterns associated with the El Niño/Southern Oscillation, *Monthly Weather Review* 115: 1606–1626.

Saji, N.H., B.N. Goswami, P.N. Vinayachandran, and T. Yamagata. 1999. A dipole mode in the tropical Indian Ocean. *Nature* 401: 360–363.

Sponberg, K. 1999. *Compendium of climatological impacts.* National Oceanic and Atmospheric Administration, Washington, DC, 62pp.

Stenseth, N.C., A. Mysterud, G. Ottersen, J.W. Hurrell, K.-S. Chan, and M. Lima. 2002. Ecological effects of climate fluctuations. *Science* 297: 1292–1296.

Timmermann, A., J. Oberhuber, A. Bacher, M. Esch, M. Latif, and E. Roeckner. 1999. Increased El Niño frequency in a climate model forced by future greenhouse warming. *Nature* 398: 694–697.

Trenberth, K.E. and T.J. Hoar. 1996. The 1990–1995 El Niño-Southern Oscillation event: longest on record, *Geophysical Research Letters* 23: 57–60.

Trenberth, K.E. and C.J. Guillemot. 1996. Physical processes involved in the 1988 drought and 1993 floods in North America, *Journal of Climate* 9: 1288–1298.

Trenberth, K.E., G.W. Branstator, D. Karoly, A. Kumar, N.-C. Lau, and C. Ropelewski. 1998. Progress during TOGA in understanding and modeling global teleconnections associated with tropical sea surface temperatures. *Journal of Geophysical Research* 103: 14291–14324.

Tudhope, A.W. et al. 2001. Variability in the El Niño-Southern Oscillation through a glacial-interglacial cycle, *Science* 291: 1511–1517.

Turner, J. 2004. The El Niño-Southern Oscillation and Antarctica. *International Journal of Climatology* 24: 1–31.

van Oldenborgh, G.J., G. Burgers, and A.K. Tank. 2000. On the El Niño teleconnection to spring precipitation in Europe. *International Journal of Climatology* 20: 565–574.

van Oldenbourgh, G.J., S.Y. Philip, and M. Collins. 2005. El Niño in a changing climate: a multi-model study. *Ocean Science* 1: 81–95.

Vecchi, G.A. and D.E. Harrison. 2000. Tropical Pacific sea surface temperature anomalies, El Niño, and equatorial westerly wind events. *Journal of Climate* 13: 1814–1830.

Wang, C. and J. Picaut. 2004. Understanding ENSO physics – a review. In: *Earth's climate: the ocean-atmosphere interaction* (eds., C. Wang, S.-P. Xie, and J.A. Carton,), Geophysical Monograph Series, pp. 21–48. AGU, Washington, DC. Webster, P.J., A.M. Moore, J.P. Loschnigg, and R.R. Leben. 1999. Coupled ocean–atmosphere dynamics in the Indian Ocean during 1997–1998. *Nature* 401: 356–360.

Wunsch, C. 1999. The interpretation of short climate records, with comments on the North Atlantic and Southern Oscillations. *Bulletin of the American Meteorogical Society* 80: 245–255.

Yuan, X. 2004. ENSO-related impacts on Antarctic sea ice: a synthesis of phenomenon and mechanisms. *Antarctic Science* 16: 415–425.

CHAPTER 2

HOW DO SCALE AND SAMPLING RESOLUTION AFFECT PERCEIVED ECOLOGICAL VARIABILITY AND REDNESS?

JOHN M. HALLEY

Department of Ecology, School of Biology, Aristotle University of Thessaloniki, U.P.B. 119, 54124 Thessaloniki, Greece
Phone +30-2310-998326, Fax: +30-2310-998379, E-mail: jmax@bio.auth.gr

2.1 ABSTRACT

Analyses of variability have been carried out on ecological time series sampled in various ways from laboratory microcosms, living populations, and populations reconstructed in various ways. The timescales of these series range from days to

David A. Vasseur and Kevin S. McCann, The Impact of Environmental Variability on Ecological Systems, 17–40. © 2007 *Springer.*

millennia. In order to use such disparate results to recreate the "big picture" of how variability depends on scale we need to interpret the results appropriately, because the sampling process itself affects the perceived variance and redness of the time series. In particular, the process associated with the collection of each data point and its timescale must be known. For example, how can we compare the variability and redness of a 50-year time series, sampled annually, with a 1,700-year time series reconstructed from sediments aggregated over 5-year intervals? In this paper, I outline some theoretical approaches and illustrate, with reference to this example, how these rules can enable us to compare very different time series.

Keywords: population variability, sampling regime, timescale, time series.

2.2 INTRODUCTION

What variability can be inferred for a biological process on the basis of a finite time series? Much of our knowledge of biological processes, often our only knowledge, comes to us through time series. The question is therefore an important one. The series in question may be environmental ones such as temperature (Engen and Saether 2000; Cyr and Cyr 2003), biological populations (Pimm and Redfearn 1988; Miramontes and Rohani 1998) or population parameters (Saether et al. 2005; Ellner 2003). It is well known that the perceived variability of an ecological time series depends on the observation parameters.

The three panels of Figure 2.1 below show the variance measured on the basis of three population time series from the GPDD (Global Population Dynamics Database, NERC Centre for Population Biology, Imperial College, 1999, www.sw.ic.ac.uk/cpb/cpb/gpdd.html) as a function of the length of the time series and as a function of the aggregation of the series into years or decades. The behaviour of the variability clearly is itself variable and strongly depends upon both of the parameters. The variance seems to be between one and two orders of magnitude higher for the aggregated series than for the yearly series. Also while variance seems to grow with observation length for the annual series of panels one and three, this stops or even gets reversed in the aggregated series.

We fully expect plenty of error in the variance estimate, so the question is whether the differences between the annual and decadal series in Figure 2.1 are due to this or whether they reflect something more systematic. It is desirable to have a theory of this variance, in order to be able to distinguish between what is due to sample error and what is due to systematic change of behaviour.

In ecological populations, the variance usually increases with observation time, a tendency that has been confirmed in a number of studies. This effect is related to the autocorrelation or "reddening" of the series. In general the redder the time series, the more pronounced will be the increase in variance with observation time. Various mechanisms and models have been proposed to explain this behaviour. These include age-structure (McArdle 1989), $1/f$ noise models (Halley 1996) and combinations of these including measurement errors (Akcakaya et al. 2003).

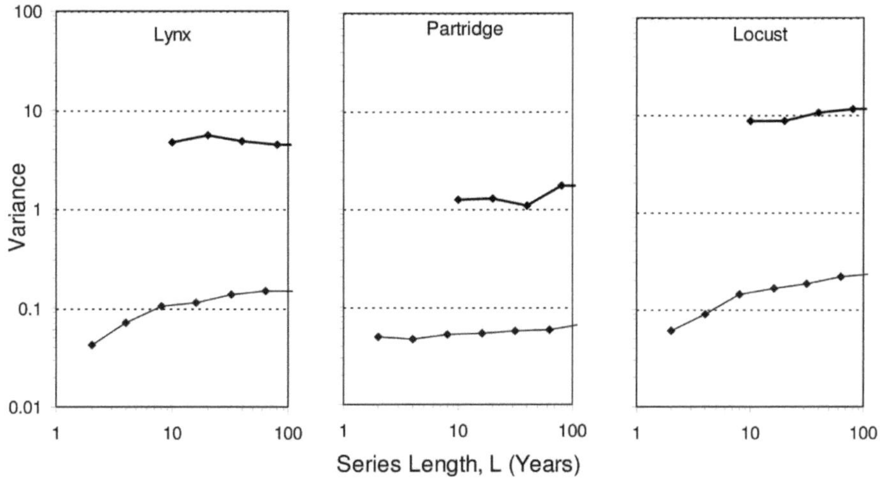

Figure 2.1. Variance of the logarithm of population for three different time series GPDD (http://cpbnts1.bio.ic.ac.uk/gpdd/). The series are (a) desert locust (*Schistocerca gregaria*), GPDD#6012, from Africa, covering 112 years, (b) grey partridge (*Perdix perdix*), GPDD#6012, from an estate in Norfolk, UK, covering 141 years (c) Canadian lynx (*Lynx Canadensis*), GPDD#453, covering 167 years. In each panel, the upper curve is the variance for total counts measured over decades while the lower curve is the variance for annual time series. At each observation length, the variance used was the average variance of a subsequence length for contiguous subsequences of the given series. (Following the methods in (Halley and Stergiou 2005)).

While the growth of variance with series length is now well appreciated, less attention has been devoted to the lower end of the observation scale, namely the timescale of each observation. Because many changes in a population happen on scales faster than the time needed to collect data for a single observation, this raises issues concerning the sampling time and the time of aggregation for each observation. That is, how often are data collected and what is the time span over which observations are aggregated into a single data point? Many models (such as a white-noise model) include implicitly all timescales down to infinitesimal sizes. That means that because the noise is effectively summed by aggregating, these invisible components contribute to the variance of each measurement. Will this issue cause problems for ecologists? A recent article by Kirchner (2005), for example, argues that for observations of $1/f$ noises, estimates of certain kinds of variability (power spectra) can be entirely misleading if these components are not explicitly accounted for. However, the same need not hold true for the measurements of variability (usually non-spectral) that are most often used in ecology. It is thus worth sorting out this issue to see how much of a problem is encountered by changing the timescales of sampling time and aggregation time.

Three timescales are likely to be involved: (a) the total series length L, (b) the sampling interval T and (c) the aggregation time τ. These are illustrated in Figure 2.2 schematically for a typical ecological time series. Note that the length L (in units

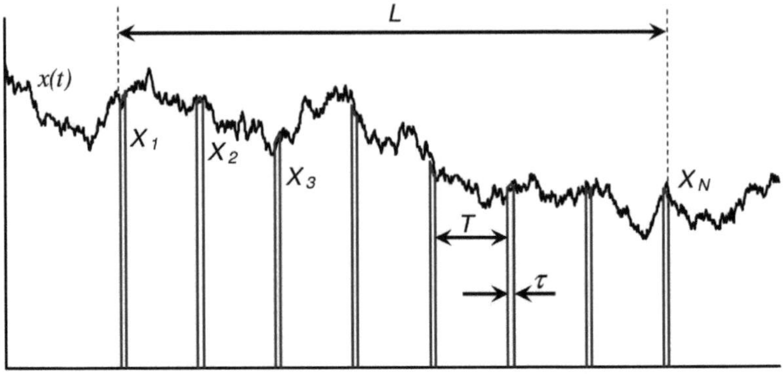

Figure 2.2. Schematic of a finite time series with imperfect sampling. The process $x(t)$ in general contains disturbances on a variety of timescales and the series is a limited window on this variability. The aggregation time τ is the duration for which the $x(t)$ is aggregated to get X_n (see Eq. (2A.3)). T is the time from the beginning of one sampling event to the next. The series duration: $L = NT - T + \tau$ is the total duration of the series from the start of the first sampling event to the end of the last which becomes $L = NT$ when $\tau =$ T.

of time) is the length of time from the start of the first sample to the end of the last sample, and is related to the number of observations N through the relation $L = (N - 1)T + \tau$. If the ecological process itself has a number of timescales, then the relation between these timescales may play a significant role in the amount of variability observed.

Apart from its intrinsic interest from a theoretical viewpoint, this is a very important issue. Firstly because we may have series collected in different ways for different timescales. An obvious example is when we wish to compare the variability of a population with its variability in the past where the latter is available only through a proxy. For example, the Baumgartner et al. (1992) time series for Pacific sardine and Northern anchovy are reconstructed from a time series of fish scales aggregated at 10-year intervals. Current population-dynamic time series are based on intervals of one year or lower. It is clear that most current time series are not long enough to distinguish between the various models (Akcakaya et al. 2003). Longer time series will be needed to address these kinds of issues, which can only be done by reconstruction of prehistoric time series by analysis of fossil surrogates, such as pollen (Clark and Mclachlan 2003) or fish scales (MacCall 1996). There are many and various problems associated with such reconstructions (e.g. Kidwell and Holland 2002; Newman and Palmer 1999). One of these problems, one that is addressed by this paper, is that the sampling and series timescales may be very different from that of contemporary time series with which we might want to compare ancient populations. This is likely to have an impact on how the past variability is understood.

The aim of this paper is to address the issue of how sampling scale affects perceived ecological variability, given that our window of observation will be a finite and imperfect time series. To answer this, a number of theoretical results are given in the appendices. Some of these are new but relevant expressions are adaptations of existing results. I draw on these results in the text in order to explain how variability depends on the timescales. In general our approach is to assume that population number is a stochastic process in continuous time. Although aware that this process is generated by process-based non-linear population dynamics of spatial extent, we will consider only the end result of the generating process, and assume that it is a Gaussian-distributed process at a point in space. This allows us see the essential dependencies of a time series arising from its characteristic scales in time. Appendix 2C lists the main symbols used in the article.

2.3 HOW THE AGGREGATION AND SAMPLING TIMES AFFECT PERCEIVED VARIABILITY

2.3.1 White noise

Suppose the process $x(t)$ in Figure 2.1 is close to white noise, namely the value at each time t is a random variable, each point independent of the previous one. Since there is no memory in the system then observation sampling time, T, will not matter. However, the aggregation time, τ, will matter. It is well known that the variance of the sum of k independent Gaussian random variables, each of variance a^2, is simply k times a^2. In the same way, Appendix 2B shows that the expected variance of the series generated by the sampling process in Figure 2.2 is:

$$V_W(\tau, T, L) = \tau a^2 \tag{2.1}$$

As a consequence of the aggregation, the expected variance is inflated by the factor τ. Thus, for a white noise process, the affect of aggregation is simply to increase the variance by the amount of aggregation. This explains why the variability, in Figure 2.1 is much higher for the aggregated series, but it does not explain why it could be nearly two orders higher. Thus, a white-noise model of variability could not explain the pattern seen in Figure 2.1. To explain these we need to turn to the issue of correlated noise processes such as the random walk.

2.3.2 The random walk

Ecological populations are known to be correlated, so that models of correlated noise are very useful. The simplest of these is the random walk (denoted "BM" hereafter since it is also called Brownian motion or brown noise), which obeys the equation:

$$\frac{dx}{dt} = bW(t) \tag{2.2}$$

$W(t)$ is Gaussian white noise with mean zero and variance unity and b is a measure of the strength of the process. Such a model has been used to model extinction, for example (Engen and Saether 2000). It is known that for a BM the variance on any stretch of the process increases linearly with its length (Schroeder 1991). Consider a continuous BM process sampled by the finite time series of Figure 2.2. The variance of this sampled process is also known to increase with series length (Arino and Pimm 1995). But how is this increase modulated by the sampling parameters? In the appendix I derive the variance of a sequence of measurements of the process $x(t)$. I show that regular sampling of the continuous BM $x(t)$ according to the scheme shown in Figure 2.2 results in a discrete BM for which variance of N successive samples is (see Appendix 2B, Eq. (2B.4)):

$$V_B(\tau, T, L) = \frac{\tau^2(N+1)(T-\tau/3)}{6}b^2 \tag{2.3}$$

Thus the sampling factor is now proportional to τ^2 rather than to τ. In the special case when observations are added across the whole sampling interval, when $T = \tau$, then (assuming also that $L \gg T$ so that $L = NT \approx (N+1)T$) the variance is:

$$V_B(\tau, \tau, L) \simeq \frac{L\tau^2}{9}b^2 \tag{2.4}$$

If on the other hand, $\tau \ll T$, (still assuming that $L \gg T$) the equation for V_B is $\tau^2 Lb^2/6$, for which the variance is still proportional to the length of the series times the square of the aggregation time. This shows that the variance is more sensitive to the timescale of aggregation if the noise is correlated, since the perceived variance of the series increases as the square of the aggregation time. Thus the variance for any such BM series aggregated at 10-year intervals should be 100 times greater than it would be if the same series were aggregated annually. This explains better the behaviour seen in Figure 2.1, where binning annual observations into 10-year observations, causes the variance to rise between one and two orders of magnitude, only now it is a little too high. For a 10-year increase in aggregation time, we expect a single order of magnitude increase of variance when the series is similar in behaviour to white noise but a 100-fold increase when the series behaves more like a BM. It is well known that ecological time series are less correlated than a random walk but more so than white noise.

2.3.3 First-order autoregressive (AR-1) processes

Ecological populations have temporal autocorrelations for various reasons, for example, finite generation time will tend to correlate the population from 1 year to the next. However, this correlation only lasts for timescales comparable to the generation time. For larger timescales, this correlation decays. For this kind of correlation, the BM is not appropriate, since correlation between the populations at two moments in time, does not decay with the time separating them. In addition, BM is characterized by a potentially unlimited increase in variance. Such "nonstationary" behaviour, where variance increases without limit, implies that there is no regulation (Murdoch and Walde 1989). A correlated but "stationary" model, one with bounded variance, is more appropriate. The simplest of such processes is the Ornstein-Uhlenbeck (OU) process (Marion et al. 2000; Gardiner 2004) which is the continuous time analogue (see Appendix 2A) of the first order autoregressive process. Indeed, as shown in Appendix 2A, finite sampling of the OU process results in the AR process:

$$X_n = \rho X_{n-1} + I_n \tag{2.5}$$

Here ρ is the correlation coefficient ($|\rho| < 1$) and I_n is a Gaussian random variable with mean zero and variance given by (2A.10). This OU process is similar to the RW process but is characterized by a specific timescale τ_c (see Eq. 2A.1) called the correlation or coherence time. The AR process in Eq. (2.5) inherits this timescale so that the correlation time can be related to the correlation coefficient ρ (defined in Appendix 2A), through the equation $\rho = \exp[-T/\tau_c]$. The correlation time, which is associated with the memory and redness of the process, marks an important boundary. When such a process is sampled after the fashion of Figure 2.2, we expect important changes in behaviour at both $T = \tau_c$ and for $\tau = \tau_c$. We give a precise description of the assumed background process in appendix 2A, wherein the expected variance V_A is also calculated. Since the formulas are quite complex, they are given in the appendix (Eq. (2A.11) or (2A.19) for an approximation). Autoregressive processes are very important in the study of correlated noise (Ripa and Lundberg 2000; Lande et al. 2003; Saether et al. 2005). As mentioned above, an important source of such correlation is age structure (McArdle 1989). Another context is the linearization of density-dependent regulation about an equilibrium (Halley and Inchausti 2004). Not only are they important in their own right, but more complicated processes can be constructed on the basis of AR processes (Steele and Henderson 1994; Halley and Kunin 1999). Figure 2.3 shows the behaviour of the series variance.

Figure 2.3a shows how the variability observed for an AR-1 process depends on the length of the time series and the aggregation time for a fixed correlation time. For very small correlation times the variance tends to increase proportional to the aggregation time for small correlation times but much faster than that for larger correlation times. Also when the correlation time is long, the variance of a time series increases with series length, L, until it reaches an asymptote. Figure 2.3b shows how the variability observed for an AR-1 process depends on the characteristic correlation time (for two cases: $\tau \ll T$ and $\tau = T$). The time series is essentially a filter that, for a given set of sampling times, sees the largest variability when the correlation time lies between the aggregation time and the series length. The variance depends upon the correlation time. If the correlation time is much shorter than the aggregation time, then only a fraction of the variability of the process is observed as so much of the variability is cancelling out due to the aggregation. On the other hand, if the correlation time is much longer than L, the time covered by the series, the variability of the process is not reflected in the time series because it cannot be detected

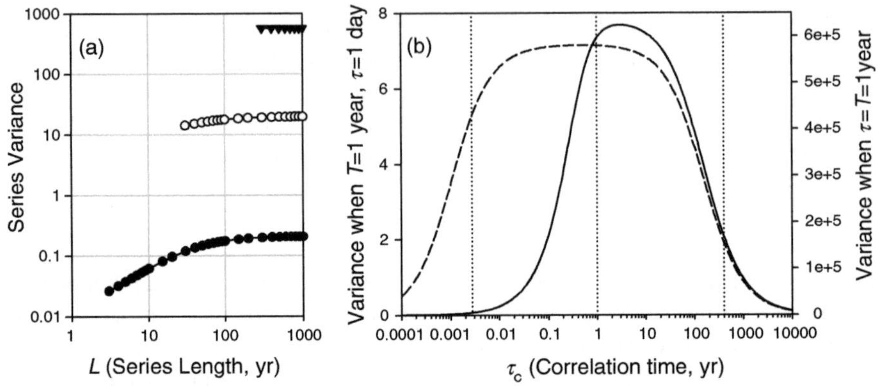

Figure 2.3. (a) The variance V_A of a correlated process seen through a finite time series according to the prescription of Figure 2.2 as a function of total series length for $T = 1(\bullet)$, $T = 10(\text{O})$ and $T = 100(\nabla)$, always with $T = \tau = 1$, and a correlation time 10 years. It is worth noting that although nearly three orders of magnitude separate the $T = \tau = 1$ curve from the $T = \tau = 10$, less than two separate the $T = \tau = 10$ curve from the $T = \tau = 100$. This latter is because for both of these cases we are in the low correlation-time regime. (b) The variance, V_A, as a function of correlation time for correlation times in the range $(10^{-4}, 10^4)$ years. The other parameters are $N = 400$, $T = 1$ year and $\sigma^2 = 3 \times 10^6$. The two curves correspond to different aggregation times: $\tau = 1$ year (continuous line) and $\tau = 1$ day (broken line). The variance was calculated using Eqs. (2A.11) with (2A.10). The important time constants associated with the time series, aggregation time τ, sampling time T and the total series time $L = NT - T + \tau$ are indicated by vertical dotted lines.

because there isn't enough time to measure it. Hence, the recommendation that the population should be sampled at or about the generation time (Connell and Sousa 1983; McArdle et al. 1990).

The approximate formula for the variance of an imperfectly sampled autoregressive process, for large N and for the special case $T = \tau$, the most important case for ecology (the more general formula is given by Eq. (2A.19) in Appendix 2A):

$$V_A(\tau, \tau, L, \tau_c) \begin{cases} \simeq 2\sigma^2 \tau_c \tau, & \tau_c \ll \tau \quad (2.6a) \\ \simeq \frac{2\sigma^2\tau^2}{3}\left[1 - \frac{2\tau_c}{L}\right], & \tau \ll \tau_c \ll L \quad (2.6b) \\ \simeq \frac{2L\tau^2}{9\tau_c}\sigma^2, & \tau_c \gg L \quad (2.6c) \end{cases}$$

This formula reflects the three zones in Figure 2.3b. Thus when the correlation time is well below the sampling time, $T = \tau$, then the sampling only picks up part of the variability because, although aggregation adds the changes, much of the fluctuations are averaged out. The variance, as for white noise, scales linearly with aggregation time. On the other hand, for large correlation time, the variance growth is proportional to L as in the case of a BM. In fact, the formula (2.6c) is the same as Eq. (2.4), $\tau_c/2$ is there to ensure that irrespective of the correlation time, the total power of the process is the same (see Appendix 2A). For intermediate times, $\tau \ll \tau_c \ll L$, the sampling factor is independent of correlation time. It is worth noting that the limit of Eq. (2.6b) for large N is not the same as for Eq. (2.6a). This is because in the case of Eq. (2.6b) the sampling time is still smaller than the correlation time and so the variability of the series is more sensitive to aggregation time. Another point to note about Eq. (2.6b) is that the approach to the asymptotic variance is very slow. It would seem natural to suppose that the difference between variance and its asymptotic value $V_A = T^2\sigma^2$ would decline exponentially as does the autocorrelation (Box et al. 1994, p. 58). However, this difference declines very slowly, as $1/L$ and this is one of the reasons why it is very difficult to distinguish an AR process from an $1/f$-noise or other "power law" noise process.

2.3.4 $1/f$-noise process

The $1/f$-noise processes (Halley 1996) is autocorrelated like Brownian motion yet less so. $1/f$-noise processes are defined by the shape of their power spectral density, their spectral power is proportional to $1/f^\nu$, where ν is called the spectral exponent. There is no specific timescale associated with any of these processes. Instead, the "redness" of these processes is determined by the magnitude of the spectral exponent. Most observed $1/f$ noises (Keshner 1982) are intermediate in redness between white noise ($\nu = 0$) and brown noise ($\nu = 2$) (but see Cuddington and Yodzis 1999). Pink $1/f$ noise has a spectral exponent of unity. This process has the special statistical and physical property that it contains equal variability at all timescales, in the sense that processes on scales of 1–9 years contribute the same as those on scales 10–90 years and so on. In practice, given the small duration of ecological time series, it is difficult to distinguish pink $1/f$ noise from an autoregressive process (Akcakaya et al. 2003) but the long-term implications are different. Unlike the AR process but like the BM process, it is characterized by infinite variance. However, this variance increases very slowly. An approximate formula is given for the special case $T = \tau$ (see Eq. (2B.11) in Appendix 2B, replacing N by L/τ):

$$V_P(\tau, \tau, L) \approx \frac{c\tau^2}{3}\left[\ln(L/\tau) + \frac{4}{3}\right] \qquad (2.7)$$

Here c is a measure of the power per octave, which is constant for a $1/f$-noise process. It is worth noting that in this case the variance is closer to the Brownian than the white-noise behaviour in that it depends on the square of aggregation time.

2.4 WHEN CAN THE VARIABILITY OF POPULATIONS BE COMPARED? AN EXAMPLE

The preceding section demonstrates that the variance calculated from a time series depends on the scales of observation, including the sampling and aggregation times. A highly desirable result would be to know how to transform into equivalence the

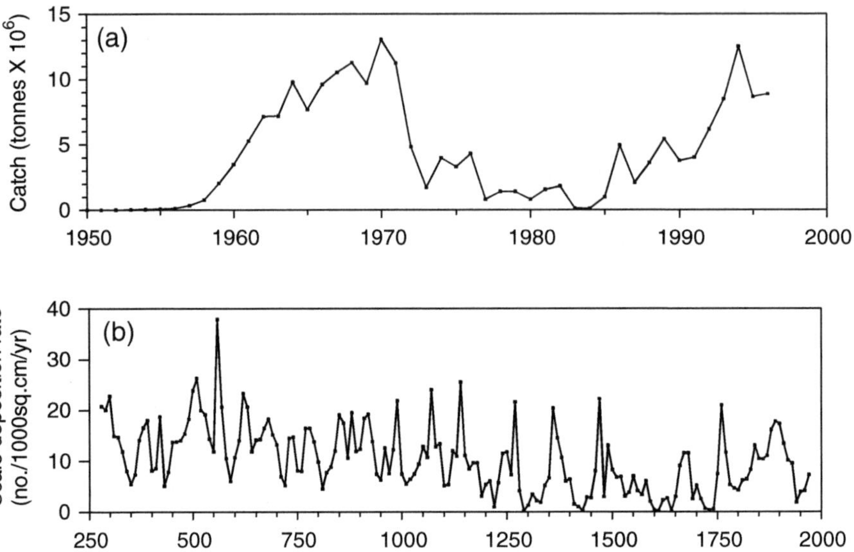

Figure 2.4. Two time series observed on different timescales. (a) The Peruvian anchovy, total yearly landings for 1950–1996, $\tau = T = 1$ year (data extracted from FAO database using FISHSTAT-PC (FAO Fisheries department, Release 1996)), (b) Time series for Northern anchovy covering the years 280–1970, reconstructed from fish time series of fish-scale-deposition rates developed from the anaerobic varved sediments of the Santa Barbara Basin off Southern California. Here, $\tau = T = 10$ years. Minor tick intervals are 50 years, the entire domain for panel (a). Without appropriate transformations, the variability of these time series cannot be compared.

variability of two time series originating from different sampling schemes. Figure 2.4 shows the time series of landings of the Peruvian anchovy (data: FAO). The total duration of the series is 47 years. Each data point represents the total annual landings for that species, so that the sampling time and the aggregation time are the same at $T = \tau = 1$ year. In the panel beneath is the reconstructed time series for Pacific Anchovy by Baumgartner et al. (1992). Each data point represents the total accumulation of fish scales at a single location. In this case, the total span of the series is over 1,700 years but the sampling time and aggregation time are both 10 years ($T = \tau = 10$).

As can be seen in Figure 2.1, for any given ecological series, subsequences of greater length, L, tend to be associated with greater variance. A matter of interest is whether this variance-growth, which is a feature of ecological populations, continues indefinitely or whether it reaches an asymptote (McArdle 1989; Cyr 1997; Inchausti and Halley 2002). In a recent paper (Halley and Stergiou 2005) we argued that the reconstructed series of Baumgartner showed this to be true as the variance of pattern of this longer series exhibits continued variance growth on millennium scales. However, there are various problems with this argument. Obviously, there are a number of ecological objections. For example, spatial variability tends to inflate the perceived temporal variance. Also note that while the FAO series are landings of the species and hence proxies for the entire population, the reconstructed time series is a proxy for the abundance of the species essentially at a single point in space (Samb and Pauly 2000). We ignore this problem because the intended scope of this paper is to identify and solve problems specifically arising from the sampling structures of series in time.

In many cases, the total variance of a process is undefined apart from the observation frame. For example, the total variance of BM and $1/f$ noise are both infinite for the limit of long series ($L \to \infty$). Thus there is no gold standard of "variability" for an ecological process obeying such models. In all models the sampling scale plays a major role in the observed variability. Figure 2.5 illustrates these issues graphically. The variance structure of Eqs. (2.1), (2.3), (2.6), and (2.7) is a surface in the space of τ, T and L. However, in this example we need only consider the simpler and highly common case where $T = \tau$ that can be represented by a three dimensional graph as shown in Figure 2.5. Essentially we have two sets of observations of variance at scales $\tau = 1$ and $\tau = 10$, for various values of L. These we call $\nu_1(L)$ and $\nu_2(L)$ respectively and we must fit these to the surface.

The fitting is complicated by the fact that there is different a multiplicative constant for each since there is a different sampling effort involved. In each of the

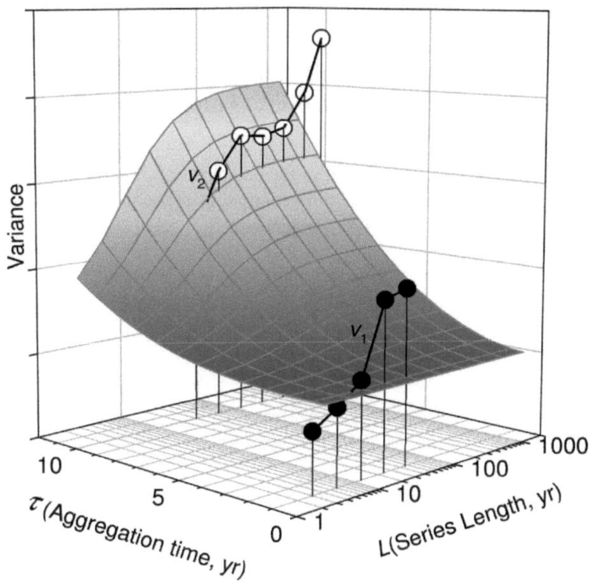

Figure 2.5. A comparison of the variance of the two series in Figure 2.4, showing the different timescales. Fitting the model to these data is a two-dimensional problem, since both the series length and the sampling/aggregation time differ. Filled circles correspond to the variance $\nu_1(L)$ of the FAO time series and open circles to the variance $\nu_2(L)$ of the reconstructed series. The surface corresponds to the variance function, $k_1 f(\tau, L)$ to which the observations are to be fitted.

different models we need to estimate a parameter corresponding to the noise intensity (which is either a^2, b^2, σ^2, or c according to the formulae (2.1), (2.3), (2.6), and (2.7) respectively) so that the general form of the model for ν_1, as a function of time series length, may be written as:

$$\nu_1(L_i) = k_1 f(\tau_1, L_i) + \varepsilon_i \tag{2.8}$$

k_1 is the noise-intensity constant (which here consists of the sampling effort times the noise intensity), while ε_i is an added noise and error term (with an expected value of zero) for the observation. The function f represents the scale-dependence or "shape factor" of the variance surface as a function of the timescales L and τ and is proportional to $V(\tau, \tau, L)$. Note that Eq. (2.8) can be treated as a regression equation, so that the noise intensity constant k_1 can be found by regressing a line ν_1 against $f(\tau_1, L)$ that passes through the origin. Similarly the model for ν_2 may be written as:

$$\nu_2(L_i) = k_2 k_1 f(\tau_2, L_i) + \eta_i \tag{2.9}$$

The extra constant k_2 is a scaling constant associated with the fact that the second series has been collected with a different form and level of sampling effort. The noise term η_i also has an expected value of zero. We calculate the scaling constant k_2 in the same way as k_1 was found, that is by regressing ν_2 against $k_1 f(\tau_2, L)$. Thus there is no need for any region of scale-overlap, which is convenient in this case since the only region of overlap is for L between 10 and 40 (see Table 2.1). We can then rescale ν_2 by k_2, namely we can plot $\nu_2(L_i)/k_2$ against the surface.

Let us consider in turn the four different types of noise model that we have analyzed to see how the associated variability is affected by sampling parameters.

Table 2.1. Statistics of the two time series used in the example. Each value of $\nu_i(L)$ was calculated by finding the average variance for all non-overlapping contiguous subsequences of length L (Inchausti and Halley 2003) both for original and time-reversed sequences

	Subseq. length, L(yr)	Variance: $\nu_i(L)$	No. Subseq. used for $\nu_i(L)$
1. FAO series			
$T = \tau = 10$ years	3	2.91	30
$L(\max) = 47$ years	6	4.72	14
$N = 47$	12	7.00	6
	24	15.19	2
Units of series = millions of tons	47	15.95	1
2. Reconstructed series			
	30	21.35	112
$T = \tau = 10$ years	60	28.37	54
$L(\max) = 1700$ years	120	32.44	26
$N = 170$	240	32.11	12
Units of series = no./100cm^2/yr	480	32.96	6
	960	37.31	2
	1700	44.20	1

2.4.1 White noise

If population variability is modelled by white noise with variance a^2, then using Eq. (2.1) in (2.8) we find that $f(L, \tau_1) = \tau_1$ and we can take the average of both sides of Eq. (8) to find $k_1 = \bar{\nu}_1/\tau_1 \approx 9.15\,(\times 10^{12}\text{tons}^2/\text{year})$. Similarly, $k_2 = \bar{\nu}_2/k_1\tau_2 = \bar{\nu}_2\tau_1/\bar{\nu}_1\tau_2 \approx 0.357$.

Dividing the variance of the second series by this constant scales the series to the fitted surface and now it is possible to see how the other points fit. The (obviously fallacious) result is shown in Figure 2.6a. The model is clearly a poor one because the variance model is constant while the data show a steady increase of variance in with L.

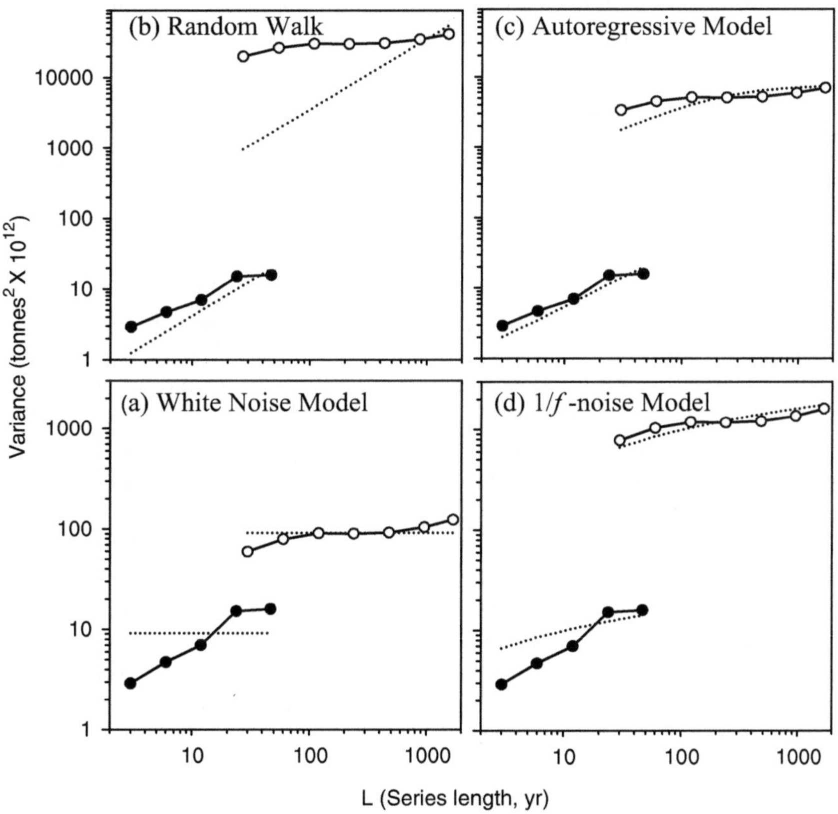

Figure 2.6. Comparison of the variance growth for the time series for Peruvian anchovy with that for the reconstructed Northern anchovy time series using the four different stochastic models of population dynamics: clockwise from bottom left: (a) white noise, (b) random walk, (c) first-order autoregressive process and (d) $1/f$−noise. The variance of the reconstructed time series $\nu_2(L)$ (open circles) has been rescaled to the units of $\nu_1(L)$, the variance of the FAO time series, (filled circles) according to the procedure outlined in the text. The dotted lines associated with series $\nu_1(L)$ and $\nu_2(L)$ represent the corresponding slice through the variance surface in Figure 2.5 (at $\tau = 1$ and $\tau = 10$ years respectively).

2.4.2 The random walk

If instead the basic population process is assumed to be a Brownian process, the length of the series matters and from Eq. (2.3) $f(L, \tau_1) = L\tau_1^2/9$, while in fitting k_1 we find the noise intensity constant b^2, fitting the regression line υ_1 as a function of f, so as to get $k_1 \approx 3.74\,(\times 10^{12}\text{tons}^2/\text{year}^3)$. The scaling constant k_2 is then derived to be $k_2 \approx 8.23 \times 10^{-4}$.

This gives a dramatically different result, as is evident from Figure 2.6b. This model does not fit so poorly to $\nu_1(L)$ but is clearly wrong for $\nu_2(L)$ where the BM predicts a much higher level of variance increase than is actually observed. Clearly the BM is also a poor model for this example of ecological variability. This underlines the observation we have already made that ecological variability tends to lie *between* white noise and the BM in terms of its variability and redness.

2.4.3 First-order autoregressive (AR-1) processes

In spite of the complex form of $f(L, \tau_1)$ given by Eq. (2.6) or (2A.11), for the case of first-order autoregressive noise, the method is exactly the same. We find k_1 by regression from Eq. (2.8).

However, in this case there is an extra parameter to find: the correlation time τ_c. If we fit AR-1 models to both series. This can be done by plotting X_{t+1} against X_t and finding the slope of the least-squares regression line through the origin as an estimate of ρ. This gives $\rho = 0.963$(per year) and $\rho = 0.869$(per decade), corresponding to time constants of 26.5 years and 71.2 years, for the fisheries landings and reconstructed time series respectively. An investigation of the compatibility of these observations from a statistical perspective is beyond the intended scope of this paper. But since these are at least the same order, we will simply estimate τ_c as the average of the two values, which gives $\tau_c = 48.9$ years.

The fitting of the observed variance to the surface, shown in Figure 2.6c, are clearly much better for this model than for either of the two preceding ones. The model captures both the initial rise of variance and also the deceleration of variance growth at longer timescales. Also, as mentioned earlier in the context of Eq. 2.6b, the approach to the asymptotic level of variance is clearly very slow, with substantial variance growth occurring even on millennial timescales much greater than the fitted correlation time of half a century.

2.4.4 $1/f$-noise process

For a $1/f$-noise model of population variability, we regress ν_1 against $\tau_1^2[\ln(L/\tau_1)+4/3]$ using the same procedure as for the other models, so as to find c. Figure 2.6d shows the fitted slices of the surface, which are also much better than the white noise or random walk models. This model captures, even better than the autoregressive model, the pattern of variance increase at longer timescales, though it is not as good a fit on the shorter timescales (because the time series in Figure 2.4a is much smoother than expected for a pink noise process and hence the variance $\nu_1(L)$ increases more slowly than for a $1/f$-noise model). In addition, this model has the advantage that no extra parameter is required.

2.5 CONCLUSIONS

Analysis of the scale-dependence of variance is one of the most fruitful ways to investigate ecological variability. With this approach, it need not be too much of a discouragement to ecologists that variability is not constant. Instead, the dependence of variability on the scale of observation becomes an important source of information about the ecological process itself. There are many challenges to us in how we interpret the results of such studies, an important one being the fact that any time series has at least two and often three timescales of measurement: the length of the series, the sampling time and the aggregation time of each sample. Most studies that have looked at the role of timescale and variability have concentrated on the first of these, mainly the more-time-more-variance effect. In this paper, I have added to this some theoretical results that also illustrate the importance of the other two timescales. How these different timescales affect the perceived variability and redness of a time series depends upon the ecological process itself.

In general the variance of an uncorrelated process is less sensitive to the observation parameters. If the ecological process is close to pure white noise, then there is no reddening and the variability is only proportional to aggregation time. On the other hand, if the process is better described by a BM or some other correlated process, then the series length and sampling time are also important. For a BM variance is proportional to the length of the series, and to the square of aggregation time. Seldom in ecology are processes very close to either of these extremes and ecological time series display behaviour intermediate between white noise and Brownian noise, having intermediate correlation and intermediate redness. The simplest intermediate model is the first-order autoregressive process, which is characterized by a single timescale, the correlation time. The perceived variance of such a process, when sampled, is highly dependent on the relationship between correlation time and the series timescales. The full equation is quite complicated but an approximate approach illustrates the different kinds of behaviour. The AR-1 process behaves like a BM process when the timescales of the series are much smaller than the correlation time and like a white noise process when, conversely, the correlation time is very small. As the length of the series increases, the variability saturates. For given series parameters and a process of fixed total power, the observed variance is greatest when the correlation time lies between the aggregation time and the total series duration. Thus a time series can be thought of as a window that only picks up certain timescales of variability. The $1/f$-noise process (also called pink noise), an alternative intermediate process between white and brown noise, is more complex than an AR-1 process but has the sometimes attractive property of being scale-free: it contains variability on all scales in equal amounts and not correlation time need be fitted.

Only some of the variability is perceived by a time series, because of the "window" effect mentioned above. The redness of the process is such that variance tends to increase as the logarithm of series length. The variance also increases approximately as the square of aggregation time, especially for long-time series. For all processes, the comparison of the variability of series sampled on different timescales requires some care.

In this paper, the analysis assumes equal bin sizes (constant τ) but we do not expect the results to be very different for unequal bin sizes provided this fluctuation is not very extreme. In our analysis we have also consistently assumed that we get the time series $X_1, X_2, \ldots, X_N$ by *aggregation* over τ of the continuous-time process $x(t)$. However, in many instances the discretization of $x(t)$ is an *averaging* rather than an aggregation. In such a case the results here for variance are simply divided by a factor τ^2 (see Appendix 2A). This means that the factor of τ becomes $1/\tau$ in Eq. (2.1) for uncorrelated noise. For the correlated models, the factor of τ^2 will drop out of Eqs. (2.3), (2.6), and (2.7), leading to a weaker dependence on τ.

These results extend our knowledge of how ecological variability can be measured using time series. There is plenty of room for further progress. Firstly, in this paper we have based our analysis on the assumption that all random variation is Gaussian. However, often population dynamics are highly skewed and this will certainly have an effect. Another issue of fundamental importance is the assumption that processes are sufficiently homogeneous in time that variance depends only on the length of series and not its position in time (this implies that there is not significant regime shifting). The process of fitting the data to the variance surface could be improved over the treatment used here where we have only used two values of the sampling time, each associated with a different scaling regime. Finally, much progress is needed on how to combine temporal and spatial scaling in a single approach to variability, but clearly the approach given here could easily be extended to a spatial context. In spite of these limitations and in spite of the considerable complexity of the systems under study, these results show many common features and illustrate the great utility of the concept of scale in studying ecological variability.

2.6 ACKNOWLEDGEMENTS

The author wishes to express thanks to an anonymous referee for useful comments and to Dr. K. Stergiou for pointing out the importance of the sampling problem in comparisons of variability.

2.7 LITERATURE CITED

Akcakaya, H.R., J.M. Halley, and P. Inchausti. 2003. Population-level mechanisms for reddened spectra in ecological time series. *Journal of Animal Ecology* 72: 698–702.

Arino, A. and S.L. Pimm. 1995. On the nature of population extremes. *Evolutionary Ecology* 9: 429–443.

Baumgartner, T.R., Soutar, A. and V. Ferreirabartrina. 1992. Reconstruction of the history of Pacific sardine and Northern anchovy populations over the past 2 millennia from sediments of the Santa-Barbara Basin, California. *California Cooperative Oceanic Fisheries Investigations Reports* 33: 24–40.

Box, G.E.P., G.M. Jenkins, and G.C. Reinsel. 1994. *Time series analysis: forecasting and control.* Prentice-Hall International, New Jersey.

Clark, J.S. and J.S. Mclachlan. 2003. Stability of forest biodiversity. *Nature* 423: 635–638.

Connell, J.H. and W.P. Sousa. 1983. On the evidence needed to judge ecological stability or persistence. *American Naturalist* 121: 789–824.

Cuddington, K.M. and P. Yodzis. 1999. Black noise and population persistence. *Proceedings of the Royal Society of London, Series B* 266: 969–973.

Cyr, H. 1997. Does inter-annual variability in population density increase with time? *Oikos* 79: 549–558.

Cyr, H, and I. Cyr. 2003. Temporal scaling of temperature variability from land to oceans. *Evolutionary Ecology Research* 5: 1183–1197.

Ellner, S.P. 2003. When does parameter drift decrease the uncertainty in extinction risk estimates? *Ecology Letters* 6: 1039–1045.

Engen, S. and B.E. Saether. 2000. Predicting the time to quasi-extinction for populations far below their carrying capacity. *Journal of Theoretical Biology* 205: 649–658.

Gardiner, C.W. *Handbook of stochastic methods*. 2004. Heidelberg, Springer.

Halley, J.M. 1996. Ecology, evolution and 1/f-noise. *Trends in Ecology and Evolution* 11: 33–37.

Halley J.M. and Inchausti P. 2004. The increasing importance of $1/f$-noises as models of ecological variability. *Fluctuation and Noise Letters* 4: R1–R26.

Halley, J.M. and W.E. Kunin. 1999. Extinction risk and the $1/f$ family of noise models. *Theoretical Population Biology* 56: 215–230.

Halley, J.M. and K.I. Stergiou. 2005. Increasing variability of fish landings. *Fish and Fisheries* 6: 266–276.

Inchausti, P. and J. Halley. 2002. The long-term temporal variability and spectral colour of animal populations. *Evolutionary Ecology Research* 4: 1033–1048.

Keshner, M.S. 1982. 1/f Noise. *Proceedings of the IEEE* 70: 212–218.

Kidwell, S.M. and S.M. Holland. 2002. The Quality of the fossil record: implications for evolutionary analyses. *Annual Review of Ecology and Systematics* 33: 561–588.

Kirchner, J.W. 2005. Aliasing in 1/F(Alpha) Noise spectra: origins, consequences, and remedies. *Physical Review E* 71, art. no. 066110.

Lande, R., S. Engen, and B.-E. Saethar. 2003. *Stochastic population dynamics in ecolgy and conservation.* Oxford University Press, Oxford.

MacCall, A.D. 1996. Patterns of low-frequency variability in fish populations of the California current. *California Cooperative Oceanic Fisheries Investigations Reports* 37: 100–110.

Marion, G., E. Renshaw, and G. Gibson. 2000. Stochastic modelling of environmental variation for biological populations. *Theoretical Population Biology* 57: 197–217.

McArdle, B. 1989. Bird population densities. *Nature* 338: 627–628.

McArdle, B.H., K.J. Gaston, and J.H. Lawton. 1990. Variation in the size of animal population: patterns, problems and artifacts. *Journal of Animal Ecology* 59: 439–454.

Miramontes, O. and P. Rohani. 1998 Intrinsically generated coloured noise in laboratory insect populations. *Proceedings of the Royal Society of London Series B, Biological Sciences* 265: 785–792.

Murdoch, W.W. and S.J. Walde. 1989. Analsyis of insect population dynamics. In: *Towards a more exact ecology* (eds., Grubb, P. and J. Whittakere.) pp. 113–140. Blackwell Scientific, Oxford.

Newman, M.E.J. and R.G. Palmer. 1999. Models of extinction: a review. *arXiv:adap-org*/9908002 v1, 49pp.

Pimm, S.L. and A. Redfearn. 1988. The variability of population-densities. *Nature* 334: 613–614.

Ripa, J. and P. Lundberg. 2000. The route to extinction in variable environments. *Oikos* 90: 89–96.

Saether, B.E., R. Lande, S. Engen, H. Weimerskirch, M. Lillegard, R. Altwegg, P.H. Becker, T. Bregnballe, J.E. Brommer, R.H. McCleery, J. Merila, E. Nyholm, W. Rendell, R.R. Robertson, P. Tryjanowski, and M.E Visser. 2005. Generation time and temporal scaling of bird population dynamics. *Nature* 436: 99–102.

Samb, B. and D. Pauly. 2000. On 'variability' as a sampling artefact: the case of Sardinella in north-western Africa. *Fish and Fisheries* 1: 206–210.

Schroeder, M. 1991. *Fractals, chaos, power laws: minutes from an infinite paradise.* W.H. Freeman, New York.

Steele, J.H. and E.W. Henderson. 1994. Coupling between physical and biological scales. *Philosophical Transactions of the Royal Society of London, Series B, Biological Sciences* 343: 5–9.

APPENDIX 2A THE SAMPLED ORNSTEIN-UHLENBECK (OU) PROCESS

In this section, I show that the OU process, a basic continuous-time process with a correlation time τ_c, becomes a first order AR process. I calculate the mean and variance of that process. This is not new, but I could not find it in standard textbooks, so I provide it here. Note that while in the text the variance is defined with respect to an argument for the total series length, L (e.g. $V_W(\tau, T, L)$ in Eq. (2.2)), here by contrast the argument is the number of points in the series N because it plays a larger role in the derivations. However, the formulae are the same.

Derivation of basic formula

Consider the OU process (Gardiner 2004, 1975):

$$\tau_c \frac{dx}{dt} + x = \sigma\sqrt{2\tau_c}W(t) \tag{2A.1}$$

Here $W(t)$ is Gaussian white noise with mean zero and variance unity. The resulting process $x(t)$ also is Gaussian with mean zero but has variance σ^2 and a correlation time of τ_c.

The Eq. (2A.1) can be integrated first with respect to time between t and $t - T$. Using the formalism of Ito calculus (Gardiner 2004) this gives:

$$x(t) - x(t-T)e^{-T/\tau_c} = \sigma\sqrt{2/\tau_c}\int_{t-T}^{t} \exp[(s-t)/\tau_c]dW(s) \tag{2A.2}$$

Now this process is sampled every T units of time by integrating x for a duration τ with the finite-time sampler leading to a time series $X_1, X_2, X_3, \ldots, X_N$. We define the sample to be:

$$X_n = \int_{nT-\tau}^{nT} x(t)dt \tag{2A.3}$$

This assumes that we get the time series $X_1, X_2, \ldots, X_N$ from $x(t)$ by aggregation. In instances where the discretization of $x(t)$ is by averaging we must include on the left-hand side a factor $1/\tau$. We also define the constant $\gamma = 1/\tau_c$ and the correlation coefficient $\rho = \exp[-T/\tau_c]$ before integrating Eq. (2A.2) again with respect to t, between nT and $nT - \tau$, using (2A.3)

$$X_n - \rho X_{n-1} = \sigma\sqrt{2\gamma} \int_{nT-\tau}^{nT} \int_{t-T}^{t} e^{\gamma(s-t)} dW(s)dt = I_n \tag{2A.4}$$

This now has the form of a first order AR process. The order of integration for the term on the right hand side can be reversed by dividing the integration into three zones, A, B, and C, as shown in Figure 2.7. Because of the stationarity of the process it suffices to do this for $n = 0$. This leads to separate integrals for zones A, B, and C:

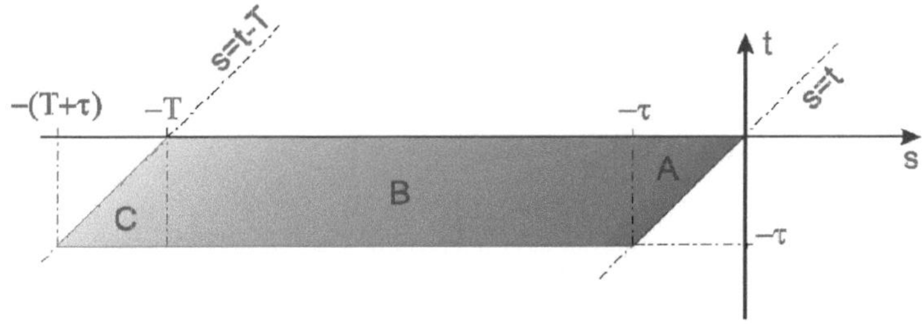

Figure 2.7. The region of integration for Eq. (2A.4). The order of integration in Eq. (2A.4) can be reversed by dividing the integration into three zones, A, B, and C, as shown. The shading is roughly proportional to the magnitude of factor $\exp[\gamma(s - t)]$ in Eq. (2A.4).

$$I_0 = I_A + I_B + I_C = \sigma\sqrt{2\gamma} \left\{ \int_{-\tau}^{0}\int_{s}^{0} + \int_{-T}^{-\tau}\int_{-\tau}^{0} + \int_{-(T+\tau)}^{-T}\int_{-\tau}^{s+T} \right\} e^{-\gamma t} dt \cdot e^{\gamma s} dW \tag{2A.5}$$

The inner integrations can be carried out to give:

$$\begin{aligned} I_0 &= \sigma\sqrt{2/\gamma} \int_{-\tau}^{0} (e^{-\gamma s} - 1) \cdot e^{\gamma s} dW + \sigma\sqrt{2/\gamma} \int_{-T}^{-\tau} (e^{\gamma\tau} - 1) \cdot e^{\gamma s} dW \\ &+ \sigma\sqrt{2/\gamma} \int_{-(T+\tau)}^{-T} (e^{\gamma\tau} - e^{-\gamma s - \gamma T}) e^{\gamma s} dW \end{aligned} \tag{2A.6}$$

We are interested only in the first two moments of I_0. The mean of I_0 is zero because $E[dW] = 0$. We can use the properties of stochastic integrals to arrive at an expression for the variance of I_0. Specifically, we use the rule (Gardiner 2004) for the expectation of the square of I_0:

$$I_0 = \int f(s) dW \quad \Rightarrow \quad E[I_0^2] = \int [f(s)]^2 ds \tag{2A.7}$$

If we square Eq. (2A.6) and take expectations, the cross terms are zero because the increments are independent. We can then apply Eq. (2A.7) to each of three remaining terms so that we are left with the integration of a sum of exponential terms that, after some algebra, yields:

$$E[I_0^2] = \frac{2\tau\sigma^2}{\gamma}\left[1 + e^{-2\gamma T}\right] - \frac{2\sigma^2}{\gamma^2}(1 - e^{-\gamma\tau})\left[1 + e^{\gamma\tau - 2\gamma T}\right] \tag{2A.8}$$

If we use the $\gamma = 1/\tau_c$ and $\rho = \exp[-T/\tau_c]$ and introduce another constant $\alpha = \exp[-\tau/\tau_c]$, then we can express Eq. (2A.8) in the more compact form (Eq. (2A.10) below, valid for all k because of the stationarity of the system).

Thus, finite imperfect sampling of the unit OU process with a correlation time τ_c results in a series of samples $X_1, X_2, X_3, \ldots, X_N$ of the AR-1 process:

$$X_n - \rho X_{n-1} = I_n \tag{2A.9}$$

With mean zero and variance σ_S^2, where:

$$\sigma_S^2 = 2\tau_c\tau\left[1 + \rho^2\right]\sigma^2 - 2\tau_c^2(1 - \alpha)\left[1 + \rho^2/\alpha\right]\sigma^2 \tag{2A.10}$$

Where the discretization of $x(t)$ is by averaging rather than aggregation Eq. (2A.10) should be scaled by a factor $1/\tau^2$. The variance of $X_1, X_2, X_3, \ldots, X_N$ is found using the standard equation for variance, which, for this AR process, is expected to have a value (McArdle 1989):

$$V_A(\tau, T, N, \tau_c) = \left[1 - 2\rho\frac{N(1-\rho) - (1-\rho^N)}{N(N-1)(1-\rho)^2}\right]\left[\frac{1}{1-\rho^2}\right]\sigma_S^2 \tag{2A.11}$$

The final estimate for V_A then is when we insert Eq. (2A.10) into Eq. (2A.11). However, this is not worth showing as the resulting expression is unwieldy and does not simplify without approximations.

Approximations

Short correlation time

When $T \geq \tau \gg \tau_c$ then both α and ρ are nearly zero, and when we use this info all the terms in brackets become unity, so the Eq. (2A.10) above becomes:

$$\sigma_S^2 \simeq 2\sigma^2\left\{\tau_c\tau - \tau_c^2\right\} \simeq 2\sigma^2\tau_c\tau$$

In this regime, the other factors in Eq. (2A.11) are easily seen to be unity since ρ is nearly zero. Thus:

$$V_A(\tau, T, N, \tau_c) \simeq 2\sigma^2\tau_c\tau \quad \forall\tau_c \ll \tau \tag{2A.12}$$

Medium correlation time

When $T \ll \tau_c \ll L$, then both $1 - \rho \ll 1$ and $1 - \alpha \ll s1$, so we expand both $\rho = \exp[-\gamma T]$ and $\alpha = \exp[-\gamma \tau]$ in their Taylor series in up to order two in γ so that Eq. (2A.10) becomes:

$$\sigma_S^2 \simeq 2\sigma^2 \tau_c \left\{ \tau \left[2 - 2\gamma T + 2\gamma^2 T^2 \right] - \tau_c \left(\gamma \tau - \frac{1}{2}\gamma^2 \tau^2 \right) \left[2 - \gamma (2T - \tau) + \frac{1}{2}\gamma^2 (2T - \tau)^2 \right] \right\}$$

When we expand the $(2T - \tau)^2$ term, and multiply out the factors then all terms in γ^0 and γ cancel, leaving only terms in γ^2 or higher:

$$\sigma_S^2 \simeq 2\sigma^2 \tau_c \tau \left[\gamma^2 T \tau - \frac{1}{3}\gamma^2 \tau^2 + O(\gamma^2) \right] \simeq \frac{2\sigma^2 \tau^2 (T - \tau/3)}{\tau_c} \tag{2A.13}$$

For the first factor in Eq. (2A.11) note that for very large N the term ρ^N is $\exp[-\gamma NT] \approx 0$ and also $N \approx N - 1$, which simplifies the factors

$$1-2\rho \frac{N(1-\rho) - (1-\rho^N)}{N(N-1)(1-\rho)^2} \simeq 1 - \frac{2\rho}{N(1-\rho)} + \frac{2\rho}{N^2(1-\rho)^2} \simeq 1 - \frac{2}{\gamma NT} \tag{2A.14}$$

For the second line we used the fact that $\rho = \exp[-\gamma T] \approx 1 - \gamma T$ and drop terms involving $1/N^2$.

The remaining factor simplifies to:

$$\frac{1}{1-\rho^2} \simeq \frac{1}{1-(1-2\gamma T)} = \frac{1}{2\gamma T} \tag{2A.15}$$

Now we can substitute Eqs. (2A.13)–(2A.15) into Eq. (2A.11) to get:

$$V_A(\tau, T, N, \tau_c) \simeq \sigma^2 \tau^2 \left[1 - \frac{\tau}{3T} \right] \left[1 - \frac{2\tau_c}{TN} \right] \qquad \forall T \ll \tau_c \ll NT \tag{2A.16}$$

Very long correlation time

In this regime the correlation time is longer than all the series timescales: $L \ll \tau_c$

First we rearrange Eq. (2A.11) as follows:

$$V_A(\tau, T, N, \tau_c) = \left[\frac{(N+1)N(1-\rho)^2 - 2N(1-\rho) + 2\rho(1-\rho^N)}{N(N-1)(1-\rho)^2} \right] \left[\frac{1}{1-\rho^2} \right] \sigma_S^2 \tag{2A.17}$$

For this regime, we already have the approximate form for the 2nd and 3rd factors of this equation via Eqs. (2A.13) and (2A.15). Now to find the first factor we expand to 3rd order in γ the terms ρ, $1-\rho$, $1-\rho^2$, $1-\rho^N$, and $(1-\rho)^2$. If we substitute these factors into the equation and drop all terms higher than 3rd order in γ then, after the requisite tedious manipulations, we get a value for the first factor of $(N+1)T/(3\tau_c)$. When we substitute this into Eq. (2A.11) along with Eqs. (2A.13) and (2A.15) we have:

$$V_A(\tau, T, N, \tau_c) \simeq \frac{(N+1)\tau^2(T-\tau/3)}{3\tau_c}\sigma^2 \qquad \forall\tau_c \gg NT \tag{2A.18}$$

Thus the equation for the variance of a sampled unit OU process can be written as:

$$V_A(\tau, T, L, \tau_c)\begin{cases} \simeq 2\sigma^2\tau_c\tau, & \tau_c \ll T \\ \simeq \sigma^2\tau^2\left[1-\frac{\tau}{3T}\right]\left[1-\frac{2\tau_c}{TN}\right], & T \ll \tau_c \ll L \\ \simeq \frac{(N+1)\tau^2(T-\tau/3)}{3\tau_c}\sigma^2, & \tau_c \gg L \end{cases} \tag{2A.19}$$

APPENDIX 2B SAMPLED WHITE, BROWNIAN AND 1/*f*-NOISE PROCESSES

The formula for the variance can be derived from the formula for the OU process.

The sampled white noise process

The sampled white noise process:

$$x = aW(t) \tag{2B.1}$$

is a limiting case of the OU process (2A.1) with $\sigma^2 = a^2/2\tau_c$ and with $\tau_c \to 0$. Thus to find the variance of this process given finite sampling we can use Eq. (2A.12) with $a^2/2\tau_c$ in place of σ^2 and then take the limit of this as $\tau_c \to 0$:

$$V_W(\tau, T, N) = \lim_{\tau_c\to 0}\left\{V_A(\tau, T, N, \tau_c)|_{\sigma=a/\sqrt{2\tau_c}}\right\} = a^2\tau \tag{2B.2}$$

The sampled random walk process

In the same way, the stochastic differential equation for the BM process:

$$\frac{dx}{dt} = bW(t) \tag{2B.3}$$

can also be seen as a limiting case of the OU process. Eq. (2B.3) can be derived from Eq. (2A.1) by setting $\sigma^2 = \tau_c b^2/2$ and taking the limit as the correlation time goes to infinity. Thus to find the variance of the sampled RW process we can therefore use the formula (2.A18) with $\sigma^2 = \tau_c b^2/2$ and then letting $\tau_c \to \infty$:

$$V_B(\tau, T, N) = \lim_{\tau_c\to\infty}\left\{\frac{(N+1)\tau^2(T-\tau/3)}{3\tau_c}\cdot\frac{b^2\tau_c}{2}\right\} = \frac{(N+1)\tau^2(T-\tau/3)}{6}b^2 \tag{2B.4}$$

The sampled $1/f$-noise process

The formula for the variance can be derived from the fact that $1/f$-noise can be approximated by a superposition of OU processes (Halley and Kunin 1999) with correlation times equidistantly spaced on a logarithmic axis. Thus the variance of the sampled process is the sum of the contributions of these independent processes

$$V_P(\tau, T, N) \simeq \sum_{k=-\infty}^{\infty} V_A(\tau, T, N, \tau_k) \tag{2B.5}$$

In the limit, the summation becomes an integral with respect to correlation time s:

$$V_P(\tau, T, N) = \int_0^{\infty} V_A(\tau, T, N, s) g(s) ds \tag{2B.6}$$

The factor $g(s)$ is the density of components per unit of time constant. For a $1/f$-noise process, the density is constant with density c on a logarithmic scale, which implies that on a linear scale the density of components must be $1/s$:

$$g(s)ds = c \cdot d(\log s) = \frac{c}{s} ds \tag{2B.7}$$

Thus the integrand is the Eq. (2A.11) times c/s which is too complex to compute without some simplification. To get a very crude approximation, we limit ourselves to the special case $\tau = T$ and we divide the integral into three zones corresponding to the three approximations of Appendix 2A.

$$V_P(T, T, N) = c \left\{ \int_0^T \int_T^{NT} \int_{NT}^{\infty} \right\} V_A(T, T, N, s) \frac{ds}{s} \tag{2B.8}$$

Using the results of Eqs. (2A.12), (2A.16), and (2A.18) for $V_A(T, T, N, s)$ we have (assume $2\sigma^2 = 1$ since this is only a scaling factor, like c):

$$V_A(T, T, N, s) \simeq \begin{cases} sT & s \ll T \\ \frac{T^2}{3}\left[1 - \frac{2s}{TN}\right] & T \ll s \ll NT \\ \frac{(N+1)T^3}{9s} & s \gg NT \end{cases} \tag{2B.9}$$

Thus we can insert these results into Eq. (2B.8)

$$V_P(T, T, N) = c \left\{ T \int_0^T ds + \frac{T^2}{3} \int_T^{NT} \left[\frac{1}{s} - \frac{2}{TN}\right] ds + \int_{NT}^{\infty} \frac{(N+1)T^3}{9s^2} ds \right\} \tag{2B.10}$$

This simplifies to the relatively compact result:

$$V_P(T, T, N) \approx \frac{cT^2}{3}\left[\ln(N) + \frac{4}{3}\right] \tag{2B.11}$$

A similar formula is easily derived for the case of $T \gg \tau$.

APPENDIX 2C SYMBOLS USED IN TEXT AND APPENDICES

Symbol	Meaning
T	Sampling time (years)
τ	Aggregation time
N	Number of observations in the time series
L	Duration of the time series ($L = NT - T + \tau$)
$V(\tau, T, L)$	The expected variance of the time series (V_W for white noise; V_B for Brownian motion; V_P for pink $1/f$-noise)
$f(\tau, L)$	Shape-factor of variance model
k_1	Noise intensity constant
k_2	Scale conversion constant between two series sampled at different efforts
τ_c	Correlation time for a 1st order process
$x(t)$	Population size at time t
X_n	Size of the nth population sample
σ^2	Total variance of an AR-1 process
a^2, b^2, σ^2, c	Intensity of white noise, random walk, AR-1, and pink noise processes respectively
ρ	Correlation coefficient of an AR-1 process $\rho = \exp[-T/\tau_c]$
α	$\alpha = \exp[-\tau/\tau_c]$
γ	$= 1/\tau_c$

CHAPTER 3

ASSESSING THE IMPACT OF ENVIRONMENTAL VARIABILITY ON TROPHIC SYSTEMS USING AN ALLOMETRIC FREQUENCY-RESOLVED APPROACH

DAVID A. VASSEUR

Department of Biological Sciences, University of Calgary, Calgary, Alberta, Canada T2N 1N4
Phone: +1 403 220-7644, Fax: +1 403 289-9311, E-mail: dvasseur@ucalgary.ca

3.1 ABSTRACT

May (1976) suggested that populations are unlikely to be influenced by environmental fluctuations that occur more frequently than the reciprocal of their characteristic response time ($1/t_r = r$). Allometric relationships, which govern the body-size-dependence of many biological and ecological phenomena, can be used to predict this characteristic response time for a wide range of populations. However, when populations are embedded in a food web, the characteristic response time becomes a property of the food web rather than of the embedded populations. This study demonstrates that, for a range of pairings of body sizes in consumer-resource systems, the characteristic response time of the system can be entirely determined by resource body size while for other pairings, it can be predominantly determined by consumer body size. Using a recently developed framework to introduce temperature variability into consumer-resource models, this study evaluates the ability of

David A. Vasseur and Kevin S. McCann, The Impact of Environmental Variability on Ecological Systems, 41–60. © 2007 *Springer.*

the characteristic response time to predict the frequency at which populations are no longer influenced by environmental fluctuations. In contrast to May's (1976) results for an isolated population, the results demonstrate that the reciprocal of the characteristic response time often represents the frequency at which the variability of resource populations is most sensitive to environmental perturbation, suggesting that community processes must be considered when determining what scales of environmental variability are important to populations.

Keywords: population, variability, response time, frequency, environmental variability.

3.2 INTRODUCTION

Allometric relationships describe the relatedness of the body size of an organism to a wide variety of other characteristics occurring at both finer and broader scales of interest (e.g. from physiological to ecological processes). These characteristics scale as the bth power of body mass (m):

$$y = am^b \tag{3.1}$$

and there has been a great deal of literature devoted to the measurement of these scaling exponents (b) and the intercepts of these relationships (a; see Peters 1983; Calder 1996). Although there remains an ongoing debate about the ubiquitous application of scaling exponents across different taxa and processes (see Kozlowski and Konarzewski 2004), there are strong arguments to suggest that we should expect scaling exponents to occur as rational multiples of the fraction $1/4$ (West et al. 1997). Furthermore, those rates which are most important for the study of population dynamics, including per capita population growth, reproduction, and metabolism, all scale identically as $m^{0.75}$ (or as $m^{-0.25}$ per-unit-mass). The conservative nature of this exponent within and between taxa makes allometric relationships useful as a tool with which to provide general descriptions of the impact of body size on population dynamics (e.g. Yodzis and Innes 1992).

For many models of population dynamics, the population growth rate r, represents the reciprocal of a population's "characteristic response time" (t_r); the fraction $1/r$ describes the time required by a population to adjust to a perturbation introduced by a singular environmental fluctuation (e.g. climate or anthropogenic disturbance). May (1976) was first to note that environmental fluctuations occurring at frequencies higher than $1/t_r$ (or synonymously with a period smaller than t_r) should have little impact on population dynamics, because such frequencies of fluctuation do not provide adequate time for a population to adjust. Since populations are constantly bombarded by environmental fluctuations, occurring over a large range of frequencies, understanding which of these are of more or less consequence has an obvious importance for determining how environmental variability impacts populations.

One can arrive at May's (1976) conclusion through the analysis of the logistic equation

$$dN/dt = rN(1 - N/K) \tag{3.2}$$

which describes the rate of change of a population's density as a function of its density N, carrying capacity K, and intrinsic growth rate r. Under constant environmental conditions, and provided that enough time has passed since initializing the model, the population will be very near its equilibrium ($N^* = K$). However, if the carrying capacity K is constantly varied, as if driven by some periodic fluctuation in an environmental variable, the population will remain further from its equilibrium; it will track the variation in the equilibrium and express a fluctuation in its own density that is highly dependent upon the period of environmental fluctuation relative to its characteristic response time. Consider the following example which is taken from May (1976), where:

$$K(t) = K_0 + K_1 \cos(2\pi t/\tau) \tag{3.3}$$

and population dynamics are modelled using Eq. 3.2 (obviously $K_1 < K_0$ in order for the carrying capacity to remain positive). When the characteristic response time is greater than the period of environmental fluctuation ($t_r \gg \tau$) the population averages across the environmental fluctuations. However, when $t_r \ll \tau$ the population closely tracks the environmental fluctuations (Figure 3.1). Given that similar dependencies were found in models with stochastic fluctuations (e.g. Roughgarden 1975), May (1976) concluded that populations will average over the high frequency components of the environmental noise spectrum while tracking those at lower frequencies. The transition between the two behaviours occurs at a frequency on the order of r (i.e. when $t_r \approx \tau$). One way to visualize this is to vary the period of environmental forcing (τ) across a range of values and record the average distance between the population and its equilibrium (Figure 3.2). When the period (τ) is much larger than t_r this average distance is very small; when the period is only slightly larger than t_r (within an order of magnitude) the average distance grows substantially; and when the period is equal to or greater than t_r the average distance is at its maximum value. Conversely the variability of the population (measured as the coefficient of variation: CV = standard deviation/mean) is greatest when the period is much larger than t_r and least when the period is greater than t_r (Figure 3.2).

The simple relationship between the characteristic response time of a population t_r and its growth rate r allows a simple determination of the allometry of response time; since the rate of growth scales as $r \propto m^{-0.25}$, the response time must scale as $t_r \propto m^{0.25}$. Thus, populations comprised of smaller bodied individuals will have a lower response time than those comprised of larger bodied individuals. Furthermore, this relationship predicts that the frequency at which environmental fluctuations are no longer influential should be lower for larger bodied organisms – a surety given that variation at the per minute scale may be important for a population of single-celled organisms but inconsequential for one comprised of large-bodied mammals.

Despite the ease with which this prediction might be tested there have been few attempts at large-scale validation, possibly due to the wide variety of timescales that

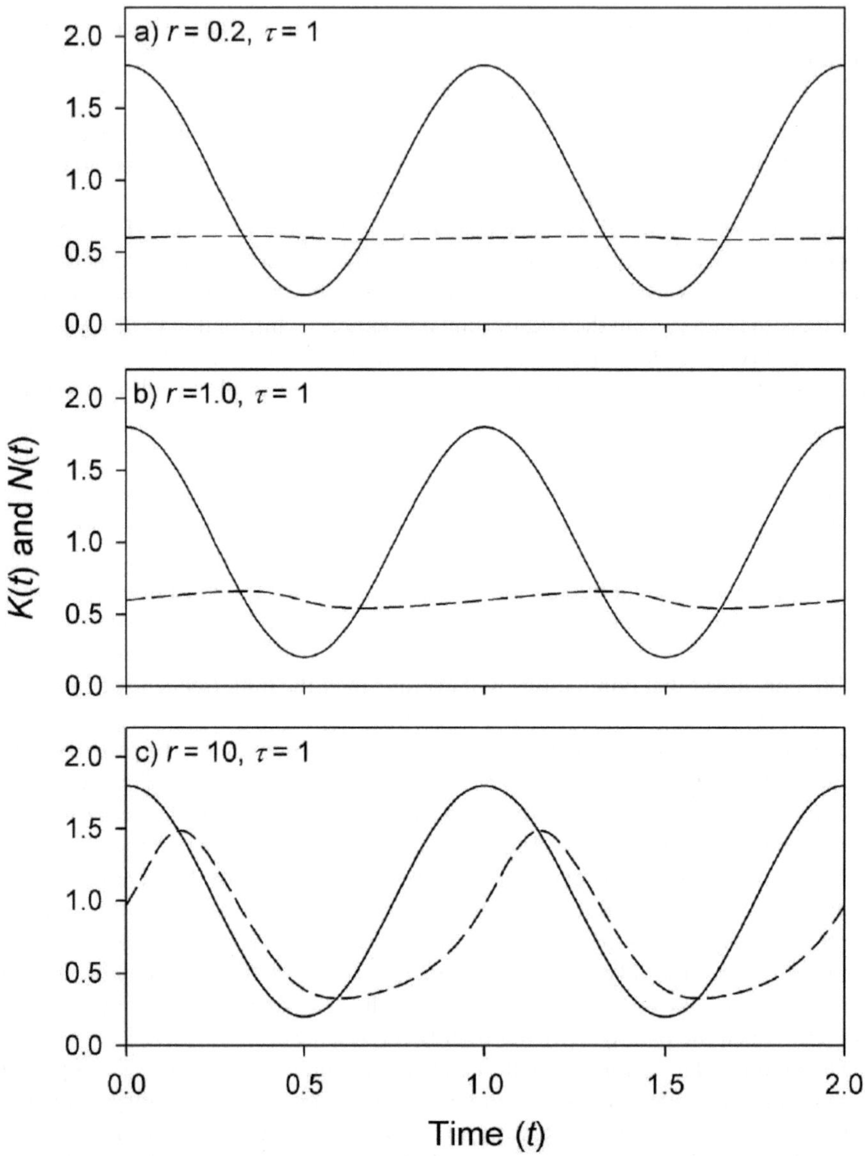

Figure 3.1. The equilibrium (solid line) and realized population density (dashed line) of the logistic model with sinusoidal variation in the carrying capacity. (From May 1976.) The three panels show examples where the frequency of environmental forcing is greater than (a), equal to (b), and less than (c) the characteristic response time.

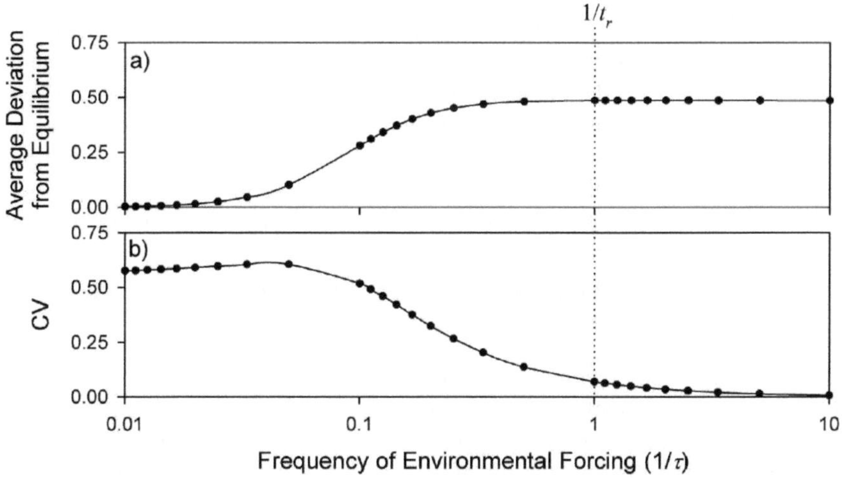

Figure 3.2. The relationship between the frequency of environmental forcing and the (a) average deviation from equilibrium (Mean Squared Error) and (b) the coefficient of variation (CV) for the logistic model with varying K (Eqs. 3.1 and 3.2). At low forcing frequencies the population density is usually near its equilibrium (cf. Figure 3.1c) and thus the mean squared error is low, while at high-forcing frequencies the population is usually far from equilibrium (cf. Figure 3.1a) resulting in a large (relative) mean squared error. The trend in the CV is opposite that of the equilibrium deviation. The dashed vertical line represents the frequency corresponding to $1/t_r$; above this frequency environmental forcing has little influence on population dynamics as evidenced by the near-zero slopes of the two curves in this region.

would be required. In perhaps the most comprehensive analysis of long-term animal population variability, Inchausti and Halley (2002) found that body size significantly influenced the spectral exponent (a measure of the expression of high relative to low frequency variability). But in contrast to theoretical predictions, they found that populations of larger-bodied primary and secondary consumers expressed more high (relative to low) frequency variation relative to their smaller-bodied counterparts (while herbivores behaved as predicted). While many factors may be responsible for the discrepancy between the allometrically based prediction and natural data, including differences in the environmental fluctuations which drive population variability and adaptations which buffer populations against environmental fluctuations, we cannot ignore the importance of ecological interactions in determining the response of populations to environmental variability. While each population in a food web has its own characteristic response time which is a function of its body size and characteristic of its internal renewal process, community processes (e.g. competition, predation) will ultimately determine a characteristic response time that is representative of the entire food web and which may depend, to varying extents, upon the body sizes and renewal processes of all the populations in the food web.

Through the analysis of a generalized allometric model for dynamic trophic systems, and using a recently developed extension of this model to include temperature, this study investigates how body size mediates the characteristic response times of small trophic communities to variability in the environment. In the first section below, I derive the relationship between body size and characteristic response time for a simple consumer-resource system using the Yodzis and Innes (1992) bioenergetic-allometric framework for dynamic trophic systems. Following this, I use numerical simulations to evaluate the utility of this measure to predict the frequencies of environmental variability that are important for population variability.

3.3 THE ALLOMETRY OF CHARACTERISTIC RESPONSE TIME IN CONSUMER-RESOURCE SYSTEMS

Yodzis and Innes (1992) pioneered the use of allometric relationships to provide logical bounds on the parameters of trophodynamic models (dynamic models of feeding relationships). For the simplest case, one consumer feeding upon one resource, they showed that although the body sizes of both populations are important determinants of the relative and total biomass in an ecological system, they do not influence the qualitative stability of the system (e.g. they have no bearing on whether the system exhibits stability, periodic oscillations, or chaos). Despite this, it has been shown that the ratio of consumer to resource body size influences the strength of interactions (Emmerson and Raffaelli 2004) and that theoretical food webs constructed from empirically plausible consumer/resource pairs, exhibit a much higher degree of stability than those with random interaction strengths (Williams and Martinez 2004; Brose and Berlow 2005). In addition to stability, body size is known to play an important role in the structuring of food webs (Cohen et al. 1993; Williams and Martinez 2000) and in their resultant dynamics (McCann and Hastings 1997; Law and Morton 1996).

The Yodzis and Innes (1992) model is a bioenergetic version of the consumer-resource system that was first proposed by Rosenzweig and MacArthur (1963). The model describes the rates of change of resource (R) and consumer (C) biomass (rather than individuals) per unit time, using the following pair of differential equations:

$$\begin{aligned} \frac{dR}{dt} &= rR\left(1-\frac{R}{K}\right) - JC\left(\frac{R}{R+R_0}\right) \\ \frac{dC}{dt} &= C\left[-M+(1-\delta)\,J\left(\frac{R}{R+R_0}\right)\right]. \end{aligned} \tag{3.4}$$

Here, resource biomass increases in the absence of consumers according to the logistic equation, which is governed by the rate of resource growth r and by the resource-carrying capacity K. The consumer population ingests resources at a rate governed by the type II functional response, where J is the maximum rate of ingestion, R_0 is the resource biomass required to realize an ingestion rate equal to $J/2$ (half-saturation density), and δ defines the fraction of biomass that is eaten but not assimilated. The consumer population loses biomass to metabolic processes at the rate M. This relatively simple model framework has three equilibria, only one of which is stable for a given set of parameters, and it can produce a variety of dynamical behaviours including stable nodes, foci, and oscillations (see Yodzis and Innes 1992; Vasseur and McCann 2005).

Body size enters the model through allometric scaling functions for the three parameters which describe per-unit-time rates of change, namely r, J, and M. Yodzis and Innes (1992) scaled these rates according to generally applicable power laws of the form of Eq. (3.1), where:

$$\begin{aligned} r &= f_r a_r m_R^{-0.25} \\ (1-\delta)J &= f_J a_J m_C^{-0.25}. \\ M &= a_M m_C^{-0.25} \end{aligned} \tag{3.5}$$

The parameters m_R and m_C represent the body sizes of resources and consumers respectively and the allometric intercepts a_r, a_J, and a_M are empirically derived constants for a "user-defined" species or set of species. Yodzis and Innes (1992) generalized these intercepts to the four metabolic classes defined by Robinson et al. (1983); endotherms, ectothermic vertebrates, ectothermic invertebrates, and unicells). Since the allometric intercepts used to represent resource growth and consumer ingestion (a_r and a_J) are usually measured under physiologically limited conditions, the fractional coefficients f_r and f_J can be used to impose ecologically limited conditions (see Yodzis and Innes 1992 for more detail).

The characteristic response time of this system can be determined, as it is for the logistic model of population growth, by calculating the eigenvalues (λ_i) of the system's Jacobian matrix. Equilibrium stability requires that all the system's eigenvalues be less than 0, but it is the most positive of the system's eigenvalues that determines the time required by the system to adjust to a change in environmental conditions. Specifically, the characteristic response time t_r is approximated by:

$$t_r = -1/\mathrm{Re}(\lambda_{\max}) \tag{3.6}$$

(Pimm and Lawton 1977). The eigenvalues of the model Eq. (3.4) and the corresponding response times are calculated for the model in Appendix 3A.

To visualize how body size influences the characteristic response time of the system, it is useful to parameterize the model and plot t_r across a range of consumer and resource body sizes. Table 3.1 provides a parameter set indicative of the interaction between a unicellular plant resource (e.g. algae) and a poikilothermic invertebrate consumer (e.g. zooplankton). The model's free parameters (R_0 and K) are chosen in the parameter space where the equilibrium is stable and approximately equidistant from the thresholds for system persistence and unstable dynamics (see Vasseur and McCann 2005). Figure 3.3 shows the characteristic response time plotted against a gradient of resource and consumer-body sizes.

Table 3.1. Model parameter descriptions and values

Parameter	Description	Value (units)	Source
Consumer-resource model			
R_0	Half-saturation density	$60\,\text{kg} \cdot \text{area}^{-1}$	
K	Carrying capacity	$10\,\text{kg} \cdot \text{area}^{-1}$	
Δ	Fraction of energy lost during ingestion and digestion	0.55	Yodzis and Innes (1992)
f_r, f_J	Realized fraction of maximum growth and ingestion rates	1	
Scaling function parameters			
a_r	Allometric intercept[1]	$0.386\,\text{kg} \cdot (\text{kg} \cdot \text{year}) \cdot \text{kg}^{0.25}$	Yodzis and Innes (1992) *in* Vasseur and McCann (2005)
a_J	Allometric intercept[1]	$9.7\,\text{kg} \cdot (\text{kg} \cdot \text{year}) \cdot \text{kg}^{0.25}$	
a_M	Allometric intercept[1]	$0.51\,\text{kg} \cdot (\text{kg} \cdot \text{year}) \cdot \text{kg}^{0.25}$	
E_r	Activation energy for resource growth	0.467 eV	Hansen et al. (1997) *in* Vasseur and McCann (2005)
E_J	Activation energy for consumer ingestion	0.772 eV	
E_M	Activation energy for consumer metabolism	0.652 eV	
k	Boltzmann's constant	$8.618\text{e}^{-5}\ \text{eV}\cdot\text{K}^{-1}$	
T	Temperature	varied (K)	
Body sizes (four sets)			
m_R	Resource body size	(A) $1.0\text{e} - 13\,\text{kg}$ (B) $1.0\text{e} - 10\,\text{kg}$ (C) $1.0\text{e} - 13\,\text{kg}$ (D) $1.0\text{e} - 10\,\text{kg}$	
m_C	Consumer body size	(A) $1.0\text{e} - 8\,\text{kg}$ (B) $1.0\text{e} - 8\,\text{kg}$ (C) $1.0\text{e} - 5\,\text{kg}$ (D) $1.0\text{e} - 5\,\text{kg}$	

[1]Allometric intercepts were measured at T_0 equals 20°C or 293 K.

There are a number of noteworthy properties that emerge from Figure 3.3. Firstly, the surface representing the characteristic response time shows two distinct zones which are separated by a vector of body-size combinations along which consumer body size is six orders of magnitude larger than resource body size ($m_C/m_R = 10^6$). This qualitative change in the system's characteristic response-time surface arises via changes in the nature of the system's eigenvalues. For combinations where $m_C/m_R > 10^6$ the system's two eigenvalues are real and distinct, generating a monotonic approach to equilibrium following perturbation. The dominant eigenvalue, and thus the characteristic response time, is strongly influenced by consumer body size, and only weakly influenced by resource body size in this region.

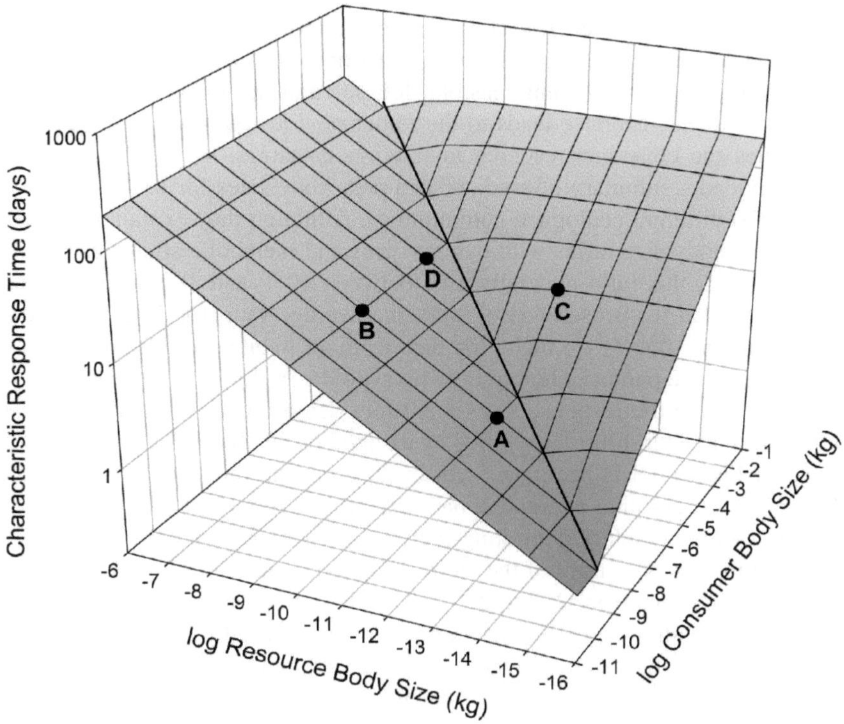

Figure 3.3. Characteristic response time surface of the body-size-dependent consumer-resource model (Eq. 3.3). The surface displays two distinct regions corresponding to different qualitative behaviours of the equilibria: when $m_C/m_R < 10^6$ the equilibrium is a focus and the response time is entirely determined by resource body size; when $m_C/m_R > 10^6$ the equilibrium is a node and the response time is determined by both resource and consumer body size. The four labelled points correspond to the four parameter sets A through D used in the analyses in later sections.

For combinations where $m_C/m_R < 10^6$ the eigenvalues are a complex conjugate pair, generating damped oscillations following perturbation. The real portion of this conjugate pair, which is used to determine the characteristic response time, is independent of consumer body size; it is determined only by the body size of the resource, through its effect on the growth rate r; $t_r \propto 1/r \propto {m_R}^{0.25}$. Although the characteristic response time of the system scales with resource body size in the same manner demonstrated by the population logistic model, the addition of a consumer to the system increases the characteristic response time by altering the intercept of the scaling function (see Appendix 3A). Pimm and Lawton (1977) were first to show that the response time of food webs quickly increases with the addition of each new trophic level to the system. This led them to the supposition that the maximum trophic level in ecological systems was limited by population dynamics rather than by the flow of energy. However, recent research has shown that decreasing the predator-prey body-size ratio at higher trophic levels can slow this increase, allowing longer chains to remain feasible (Jonsson and Ebenman 1998).

The result of this exercise leads to the rather obvious question: What body-size ratio typifies the consumer-resource interaction in natural systems? Cohen et al. (1993) provided a summary of predator and prey body sizes for 832 predator pairs from over 70 different ecological communities. Although they explain much of the variability in body-size ratios with habitat types and metabolic strategies they generally show that this ratio generally falls between $10^{1.5}$ and 10^3. In the extensive data set compiled by Brose et al. (2005), the average ratio for 16,807 consumer-resource pairs is $10^{1.74\pm3.06}$, including data from all habitats and from a variety of consumer-resource pairs (including host-parasitoid). In Figure 3.3 these ranges fall in the zone where the response time is entirely determined by resource body size and the return to equilibrium is typified by damped oscillations. However, it is worth noting that Figure 3.3 is parameterized to represent the interaction between algae and zooplankton and that alternative parameters will alter the location (but not the existence) of the boundary delimiting the two zones in Figure 3.3. Gaedke (1992) determined the size spectrum of the entire plankton community in Lake Constance, where plankton range from 2^{-6} to 2^{14} pg C and herbivorous zooplankton from 2^{13} to 2^{26} pg C. Obviously not all consumer-resource pairs falling within these ranges are valid; however, the ranges do provide conservative limits on the body-size ratios that may be possible ($10^{-0.3}$ to $10^{9.6}$). The median of this range ($10^{4.9}$) falls near the boundary in Figure 3.3 suggesting that both resource and consumer body sizes may influence the response times in natural systems.

Despite the frequency with which eigenvalues have, and continue to be used to infer the stability of an ecological system, they are an asymptotically biased measure whose error grows quickly as the system is moved further from its equilibrium. For large enough perturbations transient dynamics can be extremely important and endure for times on the order of the characteristic response time (Neubert and Caswell 1997). Although Yodzis and Innes (1992) suggested that the asymptotic approach to equilibrium appeared to be globally valid for the class of consumer-resource models described here, the characteristic response time may not adequately describe the response of the system to continued forcing at certain frequencies. For example, a perturbed system that has not returned to equilibrium before the subsequent perturbation occurs, may currently occupy a state that could excite or dampen the system's response to the subsequent perturbation. In the following sections I evaluate the ability of the characteristic response time to determine which frequencies of environmental forcing will influence population dynamics. However, before dealing with this issue, I describe some important and often overlooked assumptions which arise when considering exactly how the environment influences populations.

3.4 PARAMETRIC AND DEMOGRAPHIC PERTURBATIONS

Much of the theory surrounding the use of eigenvalues as predictors for the response of perturbed populations makes the assumption that perturbations are demographic in nature; that is, they reflect events which alter the density of the population without influencing its equilibrium, or the propensity of the population to return to its equilibrium. Such perturbations can arise from random variation among individuals in a population (e.g. quality or reproductive potential), by mass mortality events caused by extreme (rare) climatic conditions, or anthropogenic impacts. These perturbations are not directly akin to those imposed by "normal" (non-lethal) variation in environmental conditions, which is more likely to influence the parameters governing population dynamics than the population densities directly. For example the production, ingestion, and metabolic rates, which describe population dynamics in the Yodzis and Innes (1992) model are known to depend on temperature (Brown et al. 2004), and assumedly a variety of other environmental factors.

Parametric perturbations, through their effects on the parameters governing population dynamics, can indirectly influence population densities by altering the equilibrium density of the population. Recall the example presented in the introduction, where May (1976) varied the carrying capacity (K) of the logistic model through time. This perturbation influenced the population dynamics indirectly by varying the model's equilibrium; with the outcome depending upon the frequency of variation in K. However, there is an additional complication to consider when perturbing parameters – in addition to the equilibrium, the response time may itself vary with environmental conditions. Consider the dynamics of the logistic model under a new model of environmental variation which influences both r and K. Such variation will alter the equilibrium population density (K) and the characteristic response time (recall that $t_r = 1/r$) in concert. If the environmentally imposed variation in r and K is positively correlated then we expect the system to reach equilibrium faster at higher equilibrium densities. This implies that a unit increase in some environmental character (e.g. temperature, irradiance, pH) may have altogether different effects on populations than a unit decrease in the same character. Upon inspection of the equilibrium and eigenvalue equations for the Yodzis and Innes (1992) model of consumer-resource interactions (Appendix 3A) it is clear that r, K, and a number of other model parameters alter both the equilibrium and the eigenvalues of the system. This suggests that even simple model forcing (e.g. forcing of only one parameter) may produce complicated responses to perturbations, which are not predicted by determination of the eigenvalues, (and subsequently the response times) alone. Below I introduce a recent modification to the Yodzis and Innes (1992) model framework which incorporates temperature as a model parameter. Using this model I evaluate the utility of the characteristic response time to predict which frequencies of environmental forcing are important to populations.

3.5 CAN RESPONSES TIMES HELP DETERMINE HOW CONSUMERS AND RESOURCES VARY IN FLUCTUATING ENVIRONMENTS?

Any attempt to address the above question using theoretical models must make assumptions about how and where environmental variability enters the trophodynamic model. However, the risk of making such assumptions can be minimized by employing empirical relationships, which describe the dependence of model parameters, or their surrogates, on environmental conditions. Recently, the importance of temperature for a variety of biological and ecological processes including metabolic rate (Gillooly et al. 2001), developmental rate (Gillooly et al. 2002), and population growth rate (Savage et al. 2004), have been highlighted (see Brown et al. 2004 for a review). Already, these relationships have been used in theoretical models to predict temperature-induced changes in population ranges (Humphries et al. 2002), activity levels, (Humphries and Umbanhowar, Chapter 4 this volume), and energy usage (Ernest et al. 2003); and to describe gradients of global biodiversity (Allen et al. 2002) making temperature an obvious choice to address the issue at hand.

Vasseur and McCann (2005) merged the trophodynamic-allometric framework of Yodzis and Innes (1992) with the equations describing the temperature dependence of biological processes (e.g. Brown et al. 2004) to generate a trophic model that is capable of responding to environmental variability (temperature) in a mechanistic fashion. In this model, body size and temperature enter the model in the three rates r, J, and M as functions of the $-1/4$ power of resource or consumer body size (m_R, m_C) and as an exponential function of temperature (see Vasseur and McCann 2005):

$$\begin{aligned} r &= a_r(T_0)m_R^{-0.25}e^{E_r(T-T_0)/kTT_0} \\ (1-\delta)J &= a_J(T_0)m_C^{-0.25}e^{E_J(T-T_0)/kTT_0} \\ M &= a_M(T_0)m_C^{-0.25}e^{E_M(T-T_0)/kTT_0} \end{aligned} \tag{3.7}$$

where a_r, a_J,and a_M are the allometric intercepts of the body-size scaling relationships measured at the temperature T_0. The E_i are empirically estimated activation energies, k is Boltzmann's constant, and T is temperature in Kelvin. These equations follow the temperature-scaling laws developed to describe enzyme kinetics – which ultimately exert their influence upon population-level processes and are therefore suitable for biological and ecological models.

The dynamics of this model in response to long-term changes in temperature can be determined by assessing the equilibrium response to temperature. Vasseur and McCann (2005) showed that this response is governed by the differences $E_J - E_M$ and $E_r - E_M$, which control the energy requirements for the consumer and the energy budget for the system respectively. Based upon empirical estimates of the E_i, they suggested that the majority of pairings of consumers and resources would result in reduced equilibrium densities of both when the environmental temperature increased. I employ the same set of parameters used in Vasseur and McCann (2005) to describe the temperature-dependence of the interaction between unicellular algae and herbivorous zooplankton (Table 3.1) using four combinations of body sizes which fall within the range reported by Gaedke (1992) for phytoplankton and herbivorous zooplankton in Lake Constance (see Table 3.1 for parameter values).

The model provides a useful tool with which to test the influence of body size and environmental variability on the stability of consumer and resource dynamics and in addition, how the frequency of environmental forcing may influence these results. Figure 3.2 shows, using May's example for the logistic model, how the frequency of forcing influences the deviation of a population from its equilibrium values in relation to the response time. To obtain the same figures for the consumer-resource model above (Eqs. 3.4 and 3.7), the model was integrated through time, but perturbed by changes in the environmental temperature at regular intervals – to generate a specific frequency of forcing (f). Temperatures are drawn from a random normal distribution with a mean value of 20°C and a standard deviation of 5. The coefficient of variation (CV) of consumers and resources and their deviation from equilibrium conditions, were determined within each model iteration (which lasted 32,768 days) and averaged across 200 replicates at each of the frequencies tested.

For each of the four body-size combinations (Figures 3.4–3.7) it is apparent that the response of the system to different frequencies of forcing matches to some extent, the curve plotted in Figure 3.2; at low frequencies the dynamics are "at equilibrium", at intermediate frequencies the deviation from equilibrium is increasing, and at high frequencies the deviation from equilibrium is constant or decreasing. To test if May's (1976) prediction for isolated populations holds for consumer-resource pairs, the

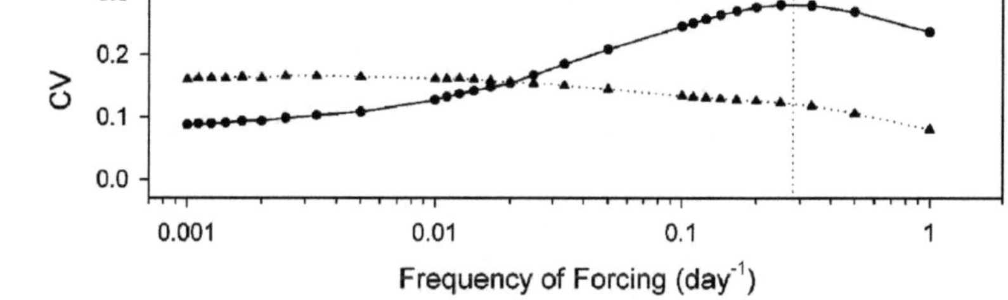

Figure 3.4. The observed relationship between the frequency of forcing and (a) the average deviation from equilibrium for resources (solid) and consumers (dotted), and (b) the CV. The vertical dotted line represents the frequency corresponding to $1/t_r$. The parameters used in this simulation are listed in Table 3.1 and body sizes are those of set A.

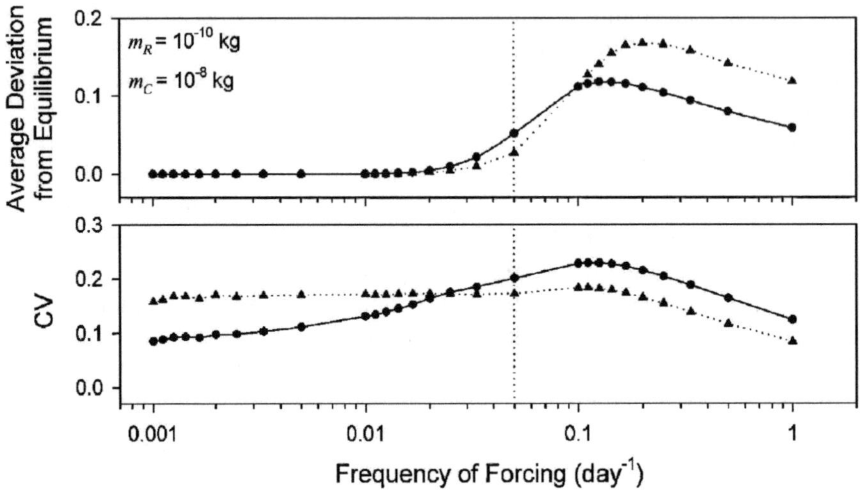

Figure 3.5. As in Figure 3.4 except body sizes are those of parameter set B.

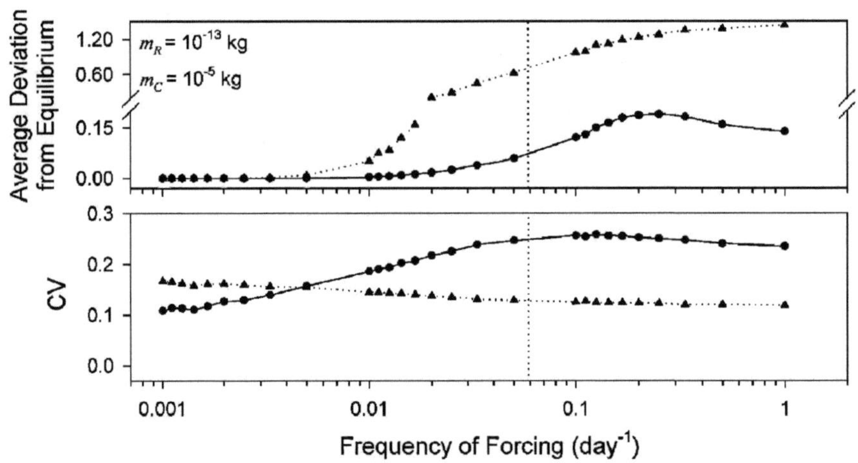

Figure 3.6. As in Figure 3.4 except body sizes are those of parameter set C.

frequency corresponding to $1/t_r$ (evaluated at 20°C) is plotted in Figures 3.3– 3.6. For all parameter sets this frequency lies within the zone where the deviation from equilibrium is increasing, suggesting that threshold frequency $1/t_r$ underestimates the importance of higher frequencies. However, there are distinct differences in the variability of the resource populations among the four body-size scenarios at, and nearby this frequency; $1/t_r$ is nearest to frequencies which produce the maximum equilibrium deviation for parameter sets A and D, but far below those for parameter sets B and C. Table 3.2 provides a summary of these results.

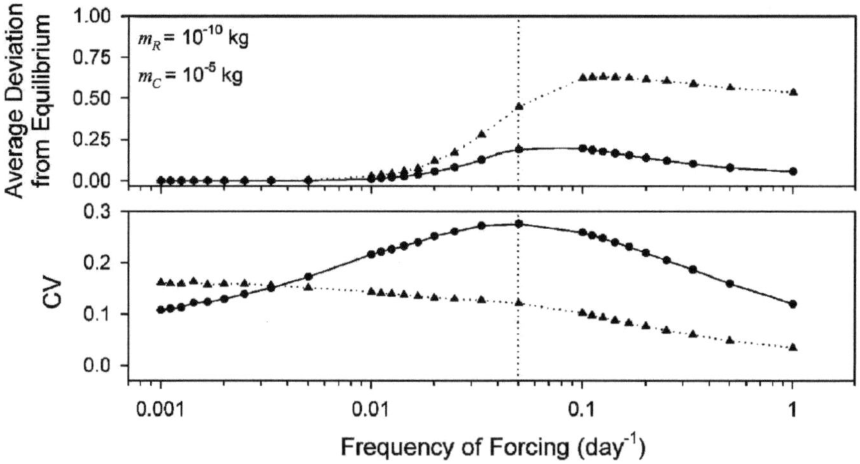

Figure 3.7. As in Figure 3.4 except body sizes are those of parameter set D.

Table 3.2. Characteristic dynamics of the four parameter sets A–D

Parameter set	Response rate ($1/t_r$) at 20°C	Largest deviation from equilibrium		Largest value of *resource* CV
		Resource	Consumer	
A	0.280 day^{-1}	0.20 @ f = 0.33 day^{-1}	0.62 @ f = 1.0 day^{-1}	0.280 @ f = 0.25 day^{-1}
B	0.050 day^{-1}	0.12 @ f = 0.13 day^{-1}	0.17 @ f = 0.20 day^{-1}	0.229@f = 0.11 day^{-1}
C	0.059 day^{-1}	0.46 @ f = 0.25 day^{-1}	3.28 @ f = 0.33 day^{-1}	0.317 @ f = 0.14 day^{-1}
D	0.050 day^{-1}	0.20 @ f = 0.10 day^{-1}	0.63 @ f = 0.13 day^{-1}	0.276 @ f = 0.05 day^{-1}

For the parameter sets A and D, the response rate $1/t_r$ denotes the (approximate) frequency at which the CV of resources is greatest and for parameter sets B and C, the peak in resource CV occurs at forcing frequencies 2–2.5 times larger than the response rate. These results are in direct contrast to those from the single population model – where the response rate $1/t_r$ denoted a frequency of forcing at and above which the population CV was relatively low (cf. Figure 3.2). The trend of having larger resource CVs at high-forcing frequencies is likely due to the ability of resources to overcompensate for changes in their equilibrium, a character that is not possible in the continuous-time logistic model. If we compare the magnitude of the equilibrium deviations and CVs among the four parameter sets, it is apparent that

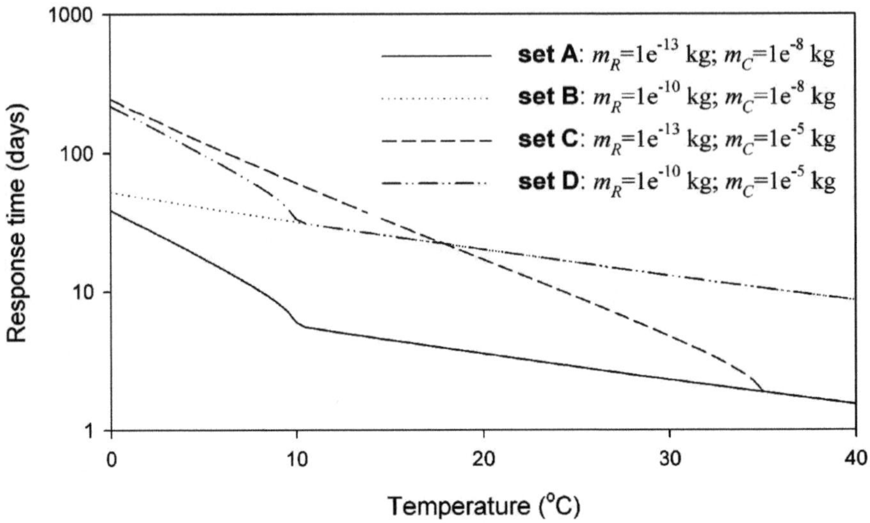

Figure 3.8. Response time of the model as a function of environmental temperature for each of the four parameter sets A through D. The curves for A, C, and D have a change in their slope corresponding to the temperature at which the system transitions from a node (at lower temperatures) to a focus (at higher temperatures).

lower absolute variability is expressed by systems whose body-size ratios are nearer to 1 (e.g. the CV for set B $<$ A $\approx$ D $<$ C).

Despite the similar response times for parameter sets B, C, and D plotted in Figure 3.1, there are evident differences in the way that these systems respond to fluctuations in environmental conditions, which arise from the underlying sensitivities of the systems' eigenvalues. As temperature varies over time, the equilibrium changes in concert with the response times of the system; at low temperatures equilibrium densities are relatively high but response times are relatively low. Figure 3.8 shows the relationship between response time and temperature for each of the four parameter sets. Over certain temperature ranges the scaling relationship between t_r and resource body size ($t_r \propto m_R{}^{0.25}$) is apparent in each curve; however, for sets A, C, and D there is a transition corresponding to a change in the fundamental properties of the equilibrium. One can envision temperature as effectively altering the body sizes of predators and prey; at low temperatures, response times are more likely to be strongly influenced by temperature and determined by both resource and consumer body size.

3.6 SUMMARY AND DISCUSSION

Characteristic response times have been cited as an important measure of the resilience of populations and trophic systems to perturbations in external conditions and given that external conditions vary continuously in most natural systems they are an important potential predictor of the level of variability these systems may express. For single populations, characteristic response times determine which frequencies of environmental variability will cause populations to themselves vary – depending on the body size of individuals in the population. Extending these results to consumer-resource models shows that when $m_C/m_R < 10^6$ response time is determined only by resource body size, and when $m_C/m_R > 10^6$ response time is mainly influenced by consumer body size (Figure 3.1). However, evaluating these results by forcing the rate parameters of the model via the Boltzmann factor shows that the characteristic response time should not be used to predict which frequencies of forcing will influence population dynamics. In fact, forcing at the frequency corresponding to $1/t_r$ often produces the largest variability one of the populations (resources).

Despite the results shown here, there are a myriad of other factors that can influence population dynamics at fixed temporal scales, or over a range of temporal scales. In practice, any population or food web will experience environmental fluctuations possessing a multitude of frequency components rather than a single frequency as I have modelled here in the latter section. While a few distinct frequencies may be responsible for the majority of environmental variation (e.g. ENSO cycles [see McPhaden, Chapter 1, this volume], seasonal cycles, lunar cycles, and diel cycles) many environmental variables are known to possess a distinct negative-scaling relationship between the frequency and magnitude of fluctuations ($f^{-\beta}$; Halley 1996; Vasseur and Yodzis 2004). This dictates that high-frequency noise has a lesser magnitude than low-frequency noise, an aspect that may lead to a reduction in resource variability at high frequencies.

Adaptive responses are also extremely important for determining the impact of environmental variability. For instance active thermoregulatory behaviours such as torpor can allow an individual to buffer environmental variability at a specific temporal scale. Humphries and Umbanhowar (Chaper 4, this volume) demonstrate that many populations traits can, under certain conditions, become decoupled from the environment through behavioural adaptations, and that this decoupling may be critical for the success of endotherms in cold climates.

Most studies examining the dynamical impacts of perturbations have remained isolated to examples involving single perturbations, in hope of being able to extrapolate their results to the more complicated cases where perturbations are not singular, but continual occurrences (Neubert and Caswell 1997). While this study provides only an example of how a consumer-resource system might respond to continually imposed environmental variability, it presents a technique whereby forcing frequencies, which excite the system can be isolated, and compared to linear measures of the forcing response (such as response time). Although I suggested earlier that communities provide a better "base" for the study of ecological variability, since populations rarely exist in isolation of others, they suffer the same criticism as they are most often embedded in larger food webs. While communities do provide a basis with which to begin understanding the importance of interspecific interactions for ecological variability, extrapolating these results to entire food webs can be highly inaccurate due to the existence of non-linear feedback mechanisms. We can project that the response times of food webs may scale predominantly with the body size of resources and that the period of forcing at which the system is most sensitive may be nearly equivalent to the response time, however, these projections are no substitute for more detailed theoretical and empirical experiments. Ultimately, ecologists need to derive a better understanding of how continual, press, and point perturbations interact to determine how ecosystems respond to environmental variability.

3.7 ACKNOWLEDGEMENTS

I would like to thank Peter Yodzis and Kevin McCann for motivating me to pursue the study of metabolic theory and its application in trophodynamic models. Murray Humphries provided valuable comments which greatly improved the manuscript. This work was partially supported by a postdoctoral fellowship from the Natural Sciences and Engineering Research Council of Canada.

3.8 LITERATURE CITED

Allen, A.P., J.H. Brown, and J.F. Gillooly. 2002. Global biodiversity, biochemical kinetics, and the energetic-equivalence rule. *Nature* 297: 1545–1548.

Brose, U. and E.L. Berlow. 2005. Scaling up keystone effects from simple to complex ecological networks. *Ecology Letters* 8: 1317–1325.

Brose, U. et al. 2005. Body sizes of consumers and their resources. *Ecology* 86: 2545.

Brown, J.H., G.B. West, and B.J. Enquist. 2000. Scaling in biology: Patterns and processes, causes and consequences. In: *Scaling in biology* (eds., J.H. Brown and G.B. West). Oxford University Press, Oxford.

Brown, J.H., J.F. Gillooly, A.P. Allen, V.M. Savage, and G.B. West. 2004. Toward a metabolic theory of ecology. *Ecology* 85: 1771–1789.

Calder, W.A. 1996. *Size, function, and life history*. Dover Publications, New York.

Cohen, J.E., S. Pimm, P. Yodzis, and J. Saldana. 1993. Body sizes of animal prey and animal predators in food webs. *Journal of Animal Ecology* 62: 67–78.

Emmerson, M.C. and D. Raffaelli. 2004. Predator-prey body size, interaction strength and the stability of a real food web. *Journal of Animal Ecology* 73: 399–409.

Ernest, S.K.M., B.J. Enquist, J.H. Brown, E.L. Charnov, J.F. Gillooly, V.M. Savage, E.P. White, F.A. Smith, E.A. Hadly, J.P. Haskell, S.K. Lyons, B.A. Maurer, K.J. Niklas, and B. Tiffney. 2003. Thermodynamic and metabolic effects on the scaling of production and population energy use. *Ecology Letters* 6: 990–995.

Gaedke, U. 1992. The size distribution of plankton biomass in a large lake and its seasonal variability. *Limnology and Oceanography* 37: 1202–1220.

Gillooly, J.F., J.H. Brown, G.B. West, V.M. Savage, and E.L. Charnov. 2001. Effects of size and temperature on metabolic rate. *Science* 293: 2248–2251.

Gillooly, J.F., E.L. Charnov, G.B. West, V.M. Savage, and J.H. Brown. 2002. Effects of size and temperature on developmental time. *Nature* 417: 70–73.

Halley, J.M. 1996. Ecology, evolution and 1/f-noise. Trends Ecol. Evol. 11: 33–37.

Hansen, P.J., P.K. Bjørnsen, and B.W. Hansen. 1997. Zooplankton grazing and growth: scaling within the 2–2,000-mm body size range. *Limnology and Oceanography* 42: 687–704.

Humphries, M.M., D.W. Thomas, and D.L. Kramer. 2002. Climate-mediated energetic constraints on the distribution of hibernating mammals. *Nature* **418**: 313–316.

Inchausti, P. and J. Halley. 2002. The long-term temporal variability and spectral colour of animal populations. Evolutionary Ecology Research 4: 1033–1048.

Jonsson, T. and B. Ebenman. 1998. Effects of predator-prey body size ratios on the stability of food chains. *Journal of theoretical biology* 193: 407–417.

Kozlowski, J. and M. Konarzewski. 2004. Is West, Brown and Enquist's model of allometric scaling mathematically correct and biologically relevant? *Functional Ecology* 18: 283–289.

Law, R. and R.D. Morton. 1996. Permanence and the assembly of ecological communities. *Ecology* 77: 762–775.

May, R.M. 1976. *Theoretical ecology, principles and applications.* Blackwell Scientific, Oxford.

McCann, K. and A. Hastings. 1997. Re-evaluating the omnivory–stability relationship in food webs. *Proceedings of the Royal Society Series B* 264: 1249–1254.

Neubert, M.G. and Caswell, H. 1997. Alternatives to resilience of measuring the responses of ecological systems to perturbations. Ecology 78: 653–665.

Peters, R.H. 1983. *The ecological implications of body size*. Cambridge University Press, New York.

Pimm, S.L. and J.H. Lawton. 1977. Number of trophic levels in ecological communities. *Nature* 268: 329–331.

Robinson, W.R., R.H. Peters, and J. Zimmermann. 1983. The effects of body size and temperature on metabolic rate of organisms. *Canadian Journal of Zoology* 61: 281–288.

Rosenzweig, M.L. and R.H. MacArthur. 1963. Graphical representation and stability conditions of predator-prey interactions. *American Naturalist* 107: 209–223.

Roughgarden, J. 1975. A simple model for population dynamics in stochastic environments. *American Naturalist* 109: 713–36.

Savage, V.M., J.F. Gillooly, J.H. Brown, G.B. West, and E.L. Charnov. 2004. Effects of body size and temperature on population growth. *American Naturalist* 163: 429–441.

Vasseur, D.A. and P. Yodzis. 2004. The color of environmental noise. *Ecology* 85: 1146–1152.

Vasseur, D.A. and K.S. McCann. 2005. A mechanistic approach for modeling temperature-dependent consumer-resource dynamics. *American Naturalist* 166: 184–198.

West, G.B., J.H. Brown, and B.J. Enquist. 1997. A general model for the origin of allometric scaling laws in biology. *Science* 276: 122–126.

Williams, R.J. and N.D. Martinez. 2000. Simple rules yield complex food webs. *Nature* 404: 180–183.

Williams, R.J. and N.D. Martinez. 2004. Stabilization of chaotic and non-permanent food-web dynamics. *European Physical Journal B* 38: 297–303.

Yodzis, P. and S. Innes. 1992. Body size and consumer resource dynamics. *American Naturalist* 139: 1151–1175.

APPENDIX 3A THE CHARACTERISTIC RESPONSE TIME OF THE TROPHODYNAMIC-ALLOMETRIC MODEL.

The Yodzis and Innes (1992) allometric model of consumer-resource dynamics is defined as:

$$\begin{aligned}\frac{dR}{dt} &= rR\left(1-\frac{R}{K}\right) - JC\left(\frac{R}{R+R_0}\right)\\ \frac{dC}{dt} &= C\left[-M+(1-\delta)\,J\left(\frac{R}{R+R_0}\right)\right]\end{aligned} \qquad (3A.1)$$

where the model rates (per-unit-time parameters) are scaled with resource and consumer body size as follows:

$$\begin{aligned} r &= f_r a_r m_R^{-0.25} \\ (1-\delta)\,J &= f_J a_J m_C^{-0.25} \\ M &= a_M m_C^{-0.25} \end{aligned} \tag{3A.2}$$

This model has three equilibrium points, two of which are trivial $[(\mathrm{R}_e, \mathrm{C}_e) = (0,0), (\mathrm{K},0)]$ and the third which is defined as:

$$\begin{aligned} R_e &= \frac{R_0}{(1-\delta)J/M - 1} \\ C_e &= \left(1 - \frac{R_e}{K}\right)\left(\frac{(1-\delta)r}{M}\right) R_e \end{aligned} . \tag{3A.3}$$

The Jacobian matrix defines the matrix of partial derivatives of Eq. (3A.1) and is defined:

$$J = \begin{bmatrix} \frac{\partial R}{\partial R} & \frac{\partial R}{\partial C} \\ \frac{\partial C}{\partial R} & \frac{\partial C}{\partial C} \end{bmatrix}. \tag{3A.4}$$

Solving for the eigenvalues of J, evaluated at Eq. (3A.3) gives:

$$\lambda_{1,2} = \frac{\left.\frac{\partial R}{\partial R}\right|_{Eq_{A3}} \pm \sqrt{\left(\left.\frac{\partial R}{\partial R}\right|_{Eq_{A3}}\right)^2 - 4\left(-\left.\frac{\partial R}{\partial C}\right|_{Eq_{A3}} \left.\frac{\partial C}{\partial R}\right|_{Eq_{A3}}\right)}}{2}. \tag{3A.5}$$

It has been shown elsewhere (Vasseur and McCann 2005) that both $\lambda_{1,2} < 0$ when $\left.\frac{\partial R}{\partial R}\right|_{Eq_3} < 0$ which is true when $R_0/K < (1-\delta)J/M - 1$. Thus, the model is always stable when the consumer's perceived resource abundance (K/R_0) is sufficiently large enough to allow the consumer's realized ingestion rate to exceed its metabolic rate.

In the range of stable parameter space the two eigenvalues can be real and distinct (corresponding to an equilibrium node) or complex conjugates (corresponding to an equilibrium focus). In the latter case a noteworthy size invariance emerges in the characteristic response time, where $t_r = -1/Re\lambda_{\max}$. Here

$$Re\lambda_{\max} = \frac{1}{2}\left.\frac{\partial R}{\partial R}\right|_{Eq_{A3}} = \frac{r}{2}\left[1 - \frac{2R_e}{K} - \left(\frac{(1-\delta)J}{M}\right)\left(1 - \frac{R_e}{K}\right)\left(\frac{R_e R_0}{(R_e + R_0)^2}\right)\right] \tag{3A.6}$$

is invariant to changes in consumer body size (m_C) since both $(1-\delta)J$ and $M \propto m_C{}^{-0.25}$. When the equilibrium is a node (two real distinct eigenvalues) $\lambda_{\max}$ is a function of both resource and consumer body sizes. The transition from focus to node is shown along a gradient of resource and consumer body sizes for a specific parameter set in Figure 3.2 and the algebraic form of this function can be found in Yodzis and Innes (1992).

CHAPTER 4

FILTERING ENVIRONMENTAL VARIABILITY: ACTIVITY OPTIMIZATION, THERMAL REFUGES, AND THE ENERGETIC RESPONSES OF ENDOTHERMS TO TEMPERATURE

MURRAY M. HUMPHRIES*

Department of Natural Resource Sciences, Macdonald Campus, McGill University, Ste-Anne-de-Bellevue, Quebec H9X 3V9, Canada
**E-mail: murray.humphries@mcgill.ca, Phone: (514) 398-7885, Fax: (514) 398-7990*

JAMES UMBANHOWAR

Department of Zoology, University of Guelph, Guelph Ontario, N1G 2W1, Canada

David A. Vasseur and Kevin S. McCann, The Impact of Environmental Variability on Ecological Systems, 61–88. © 2007 *Springer*.

4.1 ABSTRACT

Metabolic ecology may be a useful framework for understanding how animals filter environmental variability into population dynamics. We explore this possibility by examining endotherm energetic responses to subannual variability in air temperature (T_a). We first explore the mechanistic basis of endotherm metabolic responses to temperature variability based on first principles of thermal exchange. We then construct an energetic model to predict how endotherms should adaptively vary activity and refuge occupation to maximize energy gain in daily and seasonally fluctuating environments. Finally, we apply this model to a representative, highly seasonal T_a time series, to examine how temporal variation in endotherm activity and energetics conform with or diverge from temporal variation in T_a. Endotherms with access to abundant resources and high-quality refuges can be characterized by widely varying relationships between T_a and activity, intake, and expenditure, whereas net energy gain and efficiency invariably increase with increasing T_a regardless of refuge and resource access. However, the spectra of all activity and energetic traits become strongly whitened, relative to reddened subannual T_a spectra, under conditions of high resources and high-quality refuges. The capacity for endotherms to differentially filter long-period T_a fluctuations in cold and variable environments may be critical given their low-energetic gains and efficiencies in these environments.

Keywords: activity, metabolism, population dynamics, filtering, temperature.

4.2 INTRODUCTION

As a subdiscipline of ecology, physiological ecology has provided considerable insight into how ecological circumstances influence physiological traits, but has revealed much less about how these physiological responses, in turn, influence ecological processes (see recent reviews by Spicer and Gaston 1999; McNab 2002). Traditionally, physiological research has been more strongly affiliated with evolutionary theory than with ecological theory. Accordingly, much of the field's attention has focused on obtaining standardized and repeatable measures of physiological traits for use in phylogenetic comparisons. Although not unjustified, this emphasis has come at the expense of a better understanding of how physiological traits expressed by populations vary across ecological space and time, and the extent to which this variation influences the distribution and abundance of these populations.

Bioenergetics defines the process by which an organism acquires food resources from the environment and converts them into growth and reproduction (Blaxter 1989; Robbins 1993).

> One way to think of an animal is as an energy processor: it acquires energy from its environment and allocates this energy among maintenance, growth, and reproduction, with the end result that a portion of the ingested energy is converted into (somatic or reproductive) biomass. (Yodzis and Innes 1992)

Because of its central role in trophic interactions and life history allocation, energy metabolism may offer mechanistic insight into many population and community processes (Odum 1963; Sibly and Calow 1986; Peters 1991; Spicer and Gaston 1999; Kooijman 2000; DeRoos and Persson 2001; McNab 2002).

Most of the variation in standardized metabolic rates observed across a wide array of taxa can be explained by just two factors – body size and temperature. The combination of energy metabolism's mechanistic importance and empirical predictability has led to recent, renewed interest in its ecological implications.

Metabolic theory may provide a conceptual foundation for much of ecology, just as genetic theory provides a foundation for much of evolutionary biology. (Brown et al. 2004)

Proponents of metabolic theory argue that (1) metabolism is a fundamental biological rate that affects all levels of biological organization, from molecules to ecosystems, (2) metabolism varies according to universal quarter-power scaling exponents and thermal influences, and (3) these scaling laws and thermal influences can be mechanistically explained by fundamental biological, physical, and chemical constraints acting on individual organisms (Gillooly et al. 2001; Brown et al. 2004). Critics of metabolic theory dispute the generality of quarter-power exponents (Glazier 2005) and argue that proclaimed mechanistic explanations for these scaling rules are in fact difficult to quantify, empirically fitted parameters (van der Meer 2006). Interestingly, few critics of metabolic theory question the general importance of metabolism to many levels of ecological organization and analysis.

We feel that the current emphasis of metabolic theory on the explanation and application of universal scaling exponents is unnecessarily restrictive. If we seek to use metabolism as foundation to assess energy flows in ecological systems, then in addition to considering how and why standardized rates of metabolism differ across vastly different body size and temperatures, we need to consider how metabolism is actually involved in the acquisition and allocation of resources. Animal populations persist or perish in natural environments that are highly variable in time.

Understanding how environmental noise is filtered into population variability is one of the primary goals of ecology. (Greenman and Benton 2003; drawing upon Cushing et al. 1998)

In the present review, we consider the initial metabolic stages in this filtration process, focusing first on a primary source of environmental variability (temperature) and the primary mechanism by which animals respond to this variation (metabolism), then we discuss the potential relevance of these responses to population and community dynamics. Our focus on endotherms requires a broadening of the typical emphasis of metabolic ecology on standardized rates of metabolism to consider the activity, energy intake, and energy expenditure of optimizing foragers inside and outside of thermal refuges.

Two simple generalities underlie our approach and findings. First, the extent to which endotherms decouple their energetics from fluctuations in ambient temperature depends on an interaction between refuge quality and resource availability that is mediated by activity optimization. Thermal refuges create the potential for endotherms to reduce the thermal dependence of their metabolism, but realizing this potential requires resource conditions that promote prolonged occupation of refuges. Second, because optimizing foragers should seek to maximize net energy gain (intake – expenditure) rather than net energy efficiency (gain/expenditure), their behavioural responses to temperature fluctuations generate counter-intuitive energetic outcomes, frequently involving substantial increases in energy expenditure and reductions in energy efficiency to achieve small net energy gains. As a result, the energetic means by which endotherms accumulate biomass from resources may be more important in determining trophic energy flows and their persistence in variable environments than the biomass gains themselves.

4.3 MECHANISMS

4.3.1 Environmental temperature variability

Terrestrial air temperatures are characterized by ~24 h periodicity due to the rotation of Earth on its polar axis and ~365-day periodicity due to Earth's rotation around the sun (given the inclination of Earth's polar axis relative to the plane of rotation). Longer-term periodicity in surface air temperatures is imposed in some marine and terrestrial regions by large-scale oscillatory climate drivers such as the North Atlantic Oscillation (NAO) and El Nino Southern Oscillation (ENSO; McPhaden 1999; Stenseth and Mysterud 2002). The periodicity of these climatic oscillations is typically ~3–4 years with strong events occurring every ~10 years. More controversial is the potential influence of ~10-year period solar cycles on earth surface temperatures (Rind 2002).

Spectral analysis provides a useful categorization of the temporal variation present in a time series by decomposing the observed variation into a series of sinusoidal waves of different periods; a spectral density plot then provides a visual representation of the prevalence of cycles occurring at each of these different periods (Halley 1996). Investigations of the effects of noise on ecological systems typically focus on random variation above and beyond the most pronounced trends in a time series. Thus, time series are typically detrended to remove spikes associated with daily or seasonal variation to permit more detailed examination of the remaining noise. Residual variation that decomposes into a spectral density plot with slope equal to zero contains an equal mix of short and long period cycles and is classified as white noise, whereas if the slope of the spectral density plot is significantly positive, then long cycles predominate and the time series is said to be reddened (Priestley 1981). Reddened time series reflect temporal correlation in data since long period cycles make it likely that time steps in close proximity to each other will be of similar value. In contrast, white noise is a special case where a value at a given point in a time series is independent of any other value in the time series. Using monthly temperature data with seasonal variability removed, Vasseur and Yodzis (2004) found that variation in mean air temperature at inland localities approximates white noise, whereas variation in mean air temperature from coastal sites and the sea surface is characterized by a reddened spectrum (i.e. long-period cycles predominate). The reddening of temperature variation in proximity to the ocean is explained by the thermal inertia of large water bodies, which eliminates the high-frequency cycles in air temperature occurring at inland locations. This thermal inertia also reduces the amplitude of daily and seasonal temperature cycles in proximity to large water bodies. In contrast, temperature variation at inland locations is characterized by a combination of pronounced daily, annual, and multi-annual cycles interspersed by white noise (Vasseur and Yodzis 2004).

4.3.2 The biological effects of temperature

Inevitably, metabolic rates and nearly all other biochemical reaction rates increase exponentially with temperature. The influence of temperature on metabolism can be expressed most fundamentally as the thermal dependence of reaction rates as determined by the proportion of molecules that possess the energy of activation. This relationship, referred to as the Boltzmann factor or the van't Hoff–Arrhenius relation, is approximated by

$$e^{-a/kT} \tag{4.1}$$

where a is the activation energy, k is Boltzmann's constant, and T is absolute temperature in K (Brown et al. 2004). Alternatively, the thermal dependence of metabolism can be expressed as a Q_{10} value, where

$$Q_{10} = (E_1/E_2)^{\left[\frac{10}{T_2 - T_1}\right]} \tag{4.2}$$

and E_1 and E_2 are whole-animal metabolic rates measured at temperatures T_1 and T_2, respectively (McNab 2002). Most empirical estimates of Q_{10} range between 2 and 3, such that a 10°C increase in temperature generates a two- to threefold increase in metabolic rate.

4.3.3 Thermal exchange between animals and their environment

As reviewed by McNab (2002), the net thermal exchange of animals (Q_{net}, joules per hour) can be calculated as

$$Q_{\text{net}} = Q_{\text{rad}} + Q_{\text{cond}} + Q_{\text{conv}} + Q_{\text{evap}}. \tag{4.3}$$

where $Q_{\text{rad}}, Q_{\text{cond}}, Q_{\text{conv}}$, and Q_{evap} represents, respectively, radiative, conductive, convective, and evaporative thermal exchange (Box 4.1).

Box 4.1

As reviewed in McNab (2002), the net thermal exchange of animals (Q_{net}, joules per hour) can be calculated as

$$Q_{\text{net}} = Q_{\text{rad}} + Q_{\text{cond}} + Q_{\text{conv}} + Q_{\text{evap}}. \tag{4.3}$$

Q_{rad} represents radiative heat exchange, which is the process of energy exchange between distant bodies via electromagnetic waves or particles travelling at the speed of light and which can be predicted from

$$Q_{\text{rad}} = \varepsilon\sigma \mathrm{A}\left(T_s^4 - T_a^4\right) \tag{4.4}$$

where ε is the emissivity of a body, σ is the Stefan–Boltzmann constant (2.04×10^{-8} J/cm^2 h K^4), A is surface area of the body in square centimeters (cm^2), and T_s and T_a are skin temperature and ambient temperature, respectively, in Kelvin. Q_{cond} represents conductive heat exchange, which is the process of heat transfer from one molecule to another without material exchange and which can be predicted from

$$Q_{\text{cond}} = -\kappa \mathrm{A}_2 \,(T_s - T_a) \tag{4.5}$$

where κ is a thermal conductivity constant and A_2 is the surface area of contact between two bodies. Q_{conv} represents convective heat exchange, which involves the bulk movement of molecules occurring in fluids, especially at an interface with a solid and which can be predicted from

$$Q_{\text{conv}} = h_c \mathrm{A}_3 \,(T_s - T_a) \tag{4.6}$$

where h_c is the convective coefficient (expressed in J/cm^2 h°C) and A_3 is the surface area subject to convection. Frequently convective heat loss involves forced convection as a result of wind exposure and in this situation h_c tends to vary with wind velocity, v, according to $h_c = v^{0.50}$. Finally, Q_{evap} represents evaporative thermal exchange and can be predicted from

$$Q_{\text{evap}} = L\dot{E} \tag{4.7}$$

where L is the latent heat of vaporization (~2.40 kJ/g H_2O at 40°C) and $\dot{E}$(g/h) is the rate at which water is vaporized, which depends, in part, on effective surface area and the water vapour density of the external air mass (McNab 2002).

For a subset of environmental circumstances, estimation of the thermal exchange of animals can be greatly simplified (McNab 2002). The four modes of heat exchange can be combined into two terms, one for "dry" heat exchange (radiation, conductance, and convection) and the other for evaporative loss. If the majority of the gradient between body temperature and air temperature exists between the body core and its surface, then core temperature can be substituted for surface temperature. Given these simplifications, and so long as wind velocity does not influence heat loss and there is not appreciable external radiant heat load, net heat flux at all biologically relevant temperatures can be estimated from

$$Q_{\text{net}} = C'(T_b - T_a) + L\dot{E} \tag{4.8}$$

where C' is dry thermal conductance (McNab 2002). If examination of thermal exchange is further restricted to cool air temperatures, where evaporation accounts for less than 10% of heat loss, Eq. (4.8) can be further simplified to the Scholander–Irving model, where

$$Q_{\text{net}} = C(T_b - T_a) \tag{4.9}$$

and C is thermal conductance.

Thus, the thermal exchange of animals with their environment is quite complex, and we must exercise caution in using air temperature alone as an index of an animal's thermal environment (Bakken 1976; Speakman 2000; Dzialowski 2005). Nevertheless, for a subset of environmental circumstances, estimation of the thermal exchange of animals can be simplified (Box 4.1). For inactive animals experiencing cool temperatures in calm, shaded environments, the Scholander–Irving model predicts

$$Q_{\text{net}} = C(T_b - T_a), \tag{4.9}$$

where Q_{net} is net thermal exchange, T_b is body temperature, T_a is air temperature, and C is thermal conductance.

4.3.4 The ecological consequences of thermal exchange among endotherms

The defining feature of homeothermic endotherms is their use of metabolic heat to regulate their body core at a constant set-point temperature that is independent of air temperature. This means that under environmental conditions where Q_{net} is negative, maintenance of a constant body temperature requires that metabolism increases with declining air temperature along a slope that equals thermal conductance (McNab 2002). Thus,

$$Q_{\text{net}} = C(T_b - T_a) = E \tag{4.10}$$

where E is metabolic rate or energy expenditure. At warmer air temperatures there is a region referred to as the thermal neutral zone (bounded by the lower and upper critical temperatures) where metabolic rate does not vary with air temperature because small, energetically insignificant adjustments in conductance (e.g. vasodilation, piloerection, and postural changes) are sufficient to maintain a constant body temperature ($Q_{\text{net}} = 0$) despite increasing air temperature. But once air temperatures exceed the upper critical temperature, and these metabolically free adjustments to conductance are no longer sufficient to prevent Q_{net} from becoming positive, the animal must begin actively dissipating heat through panting and perspiration, which causes metabolic rate to increase. Because these responses increase heat production (i.e. contribute to the problem that it solves), the slope of the increase in metabolism above the upper critical temperature is always much steeper than the slope of the increase below the lower critical temperature. As a result of the inefficiencies of metabolic solutions to heat dissipation and because the upper critical temperature of most small endotherms occurs at quite high temperatures, most free-ranging animals with access to thermally buffered microenvironments rarely expose themselves to thermal conditions that cause their upper critical temperature to be exceeded (but see for e.g. Chappell and Bartholomew 1981). On the other hand, because the lower critical temperature of most small endotherms occurs at quite moderate temperatures, and even thermally buffered microenvironments will frequently be much colder than this, many small endotherms spend much of their lives below their lower critical temperature. Under these cool to cold conditions, Eq. (4.10) provides a reasonable approximation of an endotherm's energy expenditure, again, as long as radiant heat gain and wind-imposed convection are not significant.

Thus, from above, we have an expectation based on first-order principles of heat exchange that:

$$E = C(T_b - T_a) \tag{4.11}$$

where E is the energy expenditure or metabolic rate of an endotherm, C is its thermal conductance, T_b is its body temperature, and T_a is the prevailing air temperature or ambient temperature. Given that T_b is more or less constant in homeothermic endotherms, the general expectation is:

$$E\,\alpha - CT_a. \tag{4.12}$$

If a resting endotherm is held in a small chamber and exposed to a range of ambient temperatures below its thermoneutral zone, its metabolic rate invariably increases with declining air temperature (McNab 2002). The rate of this increase is taken to be the animal's thermal conductance. In general, thermal conductance varies inversely with body size and the quality of surface insulation (McNab 2002).

Several broad-scale empirical reviews also suggest $E\alpha - CT_a$ in nature. Recent reviews of doubly labelled water studies of field metabolic rate (FMR) have found T_a during the measurement interval (typically measured at a nearby weather station) to be among the strongest and most consistent environmental predictors of FMR in birds and mammals, with the highest levels of expenditure coinciding with the coldest ambient temperatures (Speakman 2000; Anderson and Jetz 2005). Lovegrove (2003) has also recently confirmed the long speculated correlation between basal metabolic rate (BMR) and climate (Scholander et al. 1950), with cold climate adapted endotherms again characterized by higher rates of metabolism.

The apparent generality of the negative relationship between T_a and E in nature is intriguing because there are good reasons to expect this not be the case. First, the high-energetic demands imposed by cold environments typically coincide with low-energy supply because environmental temperature and productivity are usually positively correlated. Second, all endotherms can adaptively vary behavioural activity, many have access to thermal refugia (Walsberg 1985; Huey 1991), and some can express torpor (Guppy and Withers 1999), all of which provide the opportunity for endotherms to decouple metabolism from prevailing air temperatures. For example, the energy expenditure of free-ranging North American red squirrels (*Tamiasciurus hudsonicus*) increases rather than decreases with increasing T_a during winter, because they have access to well-insulated tree nests and increase out of the nest activity on warmer winter days (Humphries et al. 2005).

4.4 MODEL

4.4.1 Model structure and assumptions

We seek to model the activity and energetics of an endotherm reacting optimally to fluctuations in T_a. We base our modelling approach on the following highly general features of endotherm energetics.

1. Acquiring resources requires activity and activity requires increased energy expenditure
2. Occupying a refuge reduces thermoregulatory requirements but prevents acquiring resources
3. Satiation resulting from prolonged intake of abundant resources generates diminishing returns on foraging effort

For the present analysis, we focus on energetic responses to subannual temperature variability, and thus seek to examine how activity and energetic traits respond to hourly, daily, and seasonal fluctuations in T_a.

The optimal behaviour of endotherms can be predicted based on the assumption that they seek to maximize net energy gain. In the simplest form,

$$G = I - E \tag{4.13}$$

where G is the net energy gain, I is the energy ingested and assimilated, and E is the energy expended (note that for simplicity of presentation, we are defining I as assimilated intake, which is equivalent to gross energy intake multiplied by an assimilation term). As long as animals do not face constraints on maximum sustainable energy expenditure, individuals should be selected to maximize net energy gain ($G = I - E$) rather than net energy efficiency ($\text{Eff} = G/E$) (Ydenberg et al. 1994). For example, an intake of 6 kJ that requires 4 kJ of expenditure is more efficient ($\text{Eff} = (6 - 4)/4 = 0.5$) than an intake of 10 kJ that requires 7 kJ of expenditure ($\text{Eff} = 0.43$), but the latter should generally be preferred because it provides a net gain of an extra 1 kJ of energy that can be usefully applied to growth or reproduction.

Locating, capturing, ingesting, and assimilating resources invariably forces an endotherm to expend additional energy above and beyond that required for thermoregulation. The energetic constraints linking activity, intake, and expenditure can be represented as:

$$G = p(i - e_a) - (1 - p)e_i \tag{4.14}$$

where p is the proportion of time spent active, i is the rate of energy intake when active, e_a the rate of expenditure when active, and e_i the rate of energy expenditure when inactive (Kam and Degen 1997). Substituting Eq. (4.11) yields,

$$\begin{aligned} G &= p[i - AC_a(T_b - T_a)] - (1 - p)[C_i(T_b - T_a)] \text{ when } T_a < T_{\text{LC}} \\ G &= p[i - A^*\text{RMR})] - (1 - p)\text{RMR when } T_{\text{UC}} > T_a > T_{\text{LC}} \end{aligned} \tag{4.15}$$

where A is a unitless activity multiplier equal to e_a/e_m, C_a is the increase in expenditure when active per unit decrease in T_a, and C_i is the increase in expenditure when inactive per unit decrease in T_a. The best thermal refuge a terrestrial endotherm can occupy is one that enables maintenance of thermoneutrality across the entire range of air temperatures that occur in its environment. The worst thermal refuge is one that has the same thermal conductance as when the animal is outside the refuge. Thus, we define refuge quality as Q, where

$$C_i = C_a - C_a Q \tag{4.16}$$

and thus $Q = 1$ is an ideal thermal refuge and $Q = 0$ is equivalent to no refuge at all. We assume hourly intake, i, increases with each hour spent active according to:

$$i_{(t+1)} = i + R \,/\, \alpha \tag{4.17}$$

where R is resource abundance and α is a scaling coefficient that defines the extent of diminishing returns on daily foraging effort (e.g. $\alpha = 1$ no diminishing returns, $\alpha = 2$, each additional hour spent foraging returns half the intake of the previous hour).

We seek to examine how p should be varied according to $T_a$in order to maximize G, given a certain R and Q, and how this optimization will, in turn, determine the effects of T_a variation on I, $E(e_a + e_i)$, and G. For present purposes, we assume T_b is a constant of 38°C and that R is a non-depletable, donor-controlled resource that remains at a constant abundance throughout the year. Although seasonal fluctuations in resources are common, this constant resource assumption permits us to focus on the all-else-being-equal effects of environmental temperature on endotherm energetics. We parameterize the model based on the well-quantified energetics of North American red squirrels, which do not hibernate and are not known to express daily or prolonged torpor (Pauls 1978a,b, Humphries and Boutin 2000; Humphries et al. 2005).

4.4.2 Modelling results

4.4.2.1 Spectral analysis of Whitehorse hourly T_a time series in comparison with other localities

We use the observational record of hourly air temperature recorded at Whitehorse, Yukon, Canada (60°42′N, 135°4′W) between January 1 and December 31, 1996 as a representative, highly seasonal time series to evaluate energetic responses to T_a fluctuations. Figure 4.1 presents a spectral analysis of the Whitehorse hourly T_a time series from January 1961 to January 2001, in comparison with other continental and coastal localities in Canada, all of which were obtained from Environment Canada (http://climate.weatheroffice.ec.gc.ca/climateData/canada_e.html). Not surprisingly, all localities are characterized by frequency spikes at daily and annual periods (and associated resonant frequencies), as well as more subtle, broad-based peaks at 3- to 4-year and decadal periods (Figure 4.2). The overall flattening of all spectra beyond a 1 year period corresponds to Vasseur and Yodzis, (2004) finding that multi-annual variation in terrestrial air temperatures approximates white noise. Within this roughly flat section of the spectral plots, the slight tendency for the two coastal communities (Churchill and Halifax) to be characterized by more positive

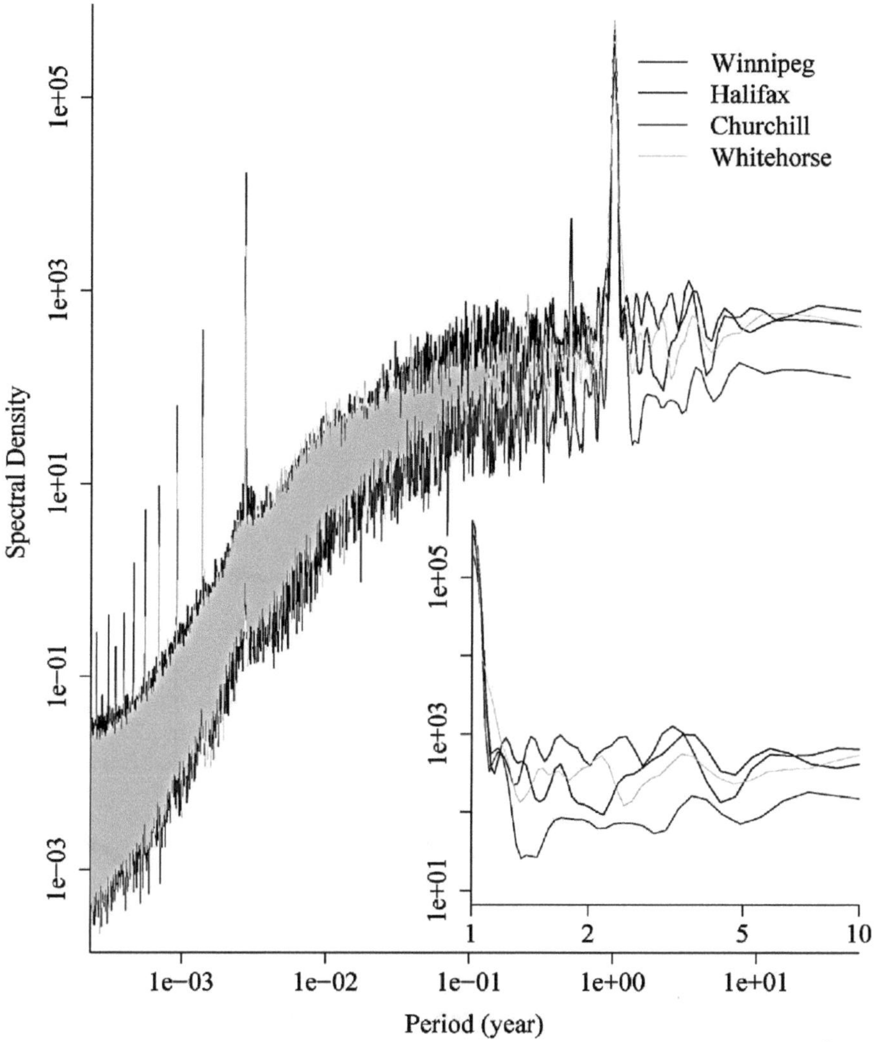

Figure 4.1. Spectral densities of hourly air temperature from 1953 to 2004 for four Canadian weather stations. Halifax, Nova Scotia (44°52′ N, 63°31′ W) and Churchill, Manitoba (58°44′, 94°3′ W) are within a few kilometers of the coast, while Whitehorse, Yukon (60°42′ N, 135°4′ W), and Winnipeg, Manitoba (49°55′ N, 97°13′ W) are located >500 km inland.

slopes than the two continental communities (Whitehorse and Winnipeg) is also consistent with Vasseur and Yodzis, (2004) conclusion that coastal sites are characterized by reddened spectra (i.e. long-period cycles predominate) relative to inland sites. However, at less than a 1-year period, which is the focus our present analysis, T_a spectra from all four localities are highly reddened. The occurrence of this reddening (and the temporal correlation in data it indicates) is not surprising given that warm and cold bouts are regular features of weather across hourly, daily, and seasonal time frames.

4.4.2.2 *Daily activity and energetics as a function of average daily T_a*

The proportion of the day spent active (p) increases with ambient temperature (T_a), but the extent and form of this increase is strongly affected by resource availability. When resources are scarce, no daily activity is expected across a wide range of low T_a, but once T_a exceeds an activity threshold, p increases rapidly with increasing T_a to a maximum activity asymptote (Figure 4.2A4–D4). When resources are abundant, daily activity is predicted at all temperatures and is expected to increase only slightly with increasing T_a because satiation eliminates the benefits of prolonged activity (Figure 4.2A1–D1). At intermediate resource levels, daily activity tends to increase curvilinearly with ambient temperature (Figure 4.2A2–D2, A3–D3). Daily

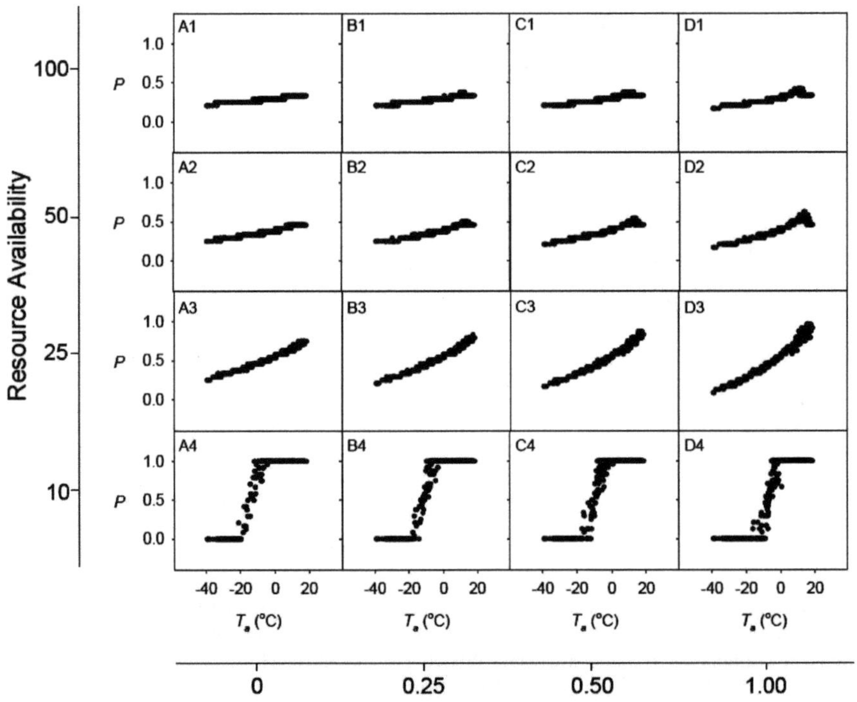

Figure 4.2. The effect of resource availability (R) and refuge quality (Q) on the relationship between ambient temperature (T_a) and daily activity (p; proportion of 24 h spent active) for an endotherm maximizing energy gain. Parameter values correspond to the well-quantified energetics of Yukon red squirrels (T_b = 38°C, C_a = C_i = 0.1917 kJ/h/°C, RMR = 4.375 kJ/h, A = 2). Refuge quality scenarios considered include Q = 0, 0.25, 0.50, and 1.0 (see Eq. (4.16). Resource scenarios considered include R = 10 and α = 1.00; R = 25 and α = 1.10; R = 50 and α = 1.25; R = 100 and α = 1.50 (see Eq. (4.17)).

activity patterns are little influenced by refuge quality because the energetic advantages of foraging activity are dictated by the resource gains and thermoregulatory expenditures experienced away from the refuge. A high-quality refuge cannot deter animals from foraging in profitable environments because net gain while in the refuge is always negative, even for the highest quality refuge. However, when foraging activity offers negative energy returns that approximate the negative returns offered by remaining inactive in a refuge, a higher quality refuge causes animals to await warmer air temperatures before initiating activity (compare activity thresholds in Figure 4.2A4–D4).

The expected relationship between energy intake (I) and T_a generally mirrors the expected relationship between activity and T_a because time spent active is a prerequisite for energy intake. Accordingly, the relationship between I and T_a is little affected by refuge quality but strongly affected by resource availability (Figure 4.3, light gray symbols). When resources are scarce, no intake is predicted across a wide

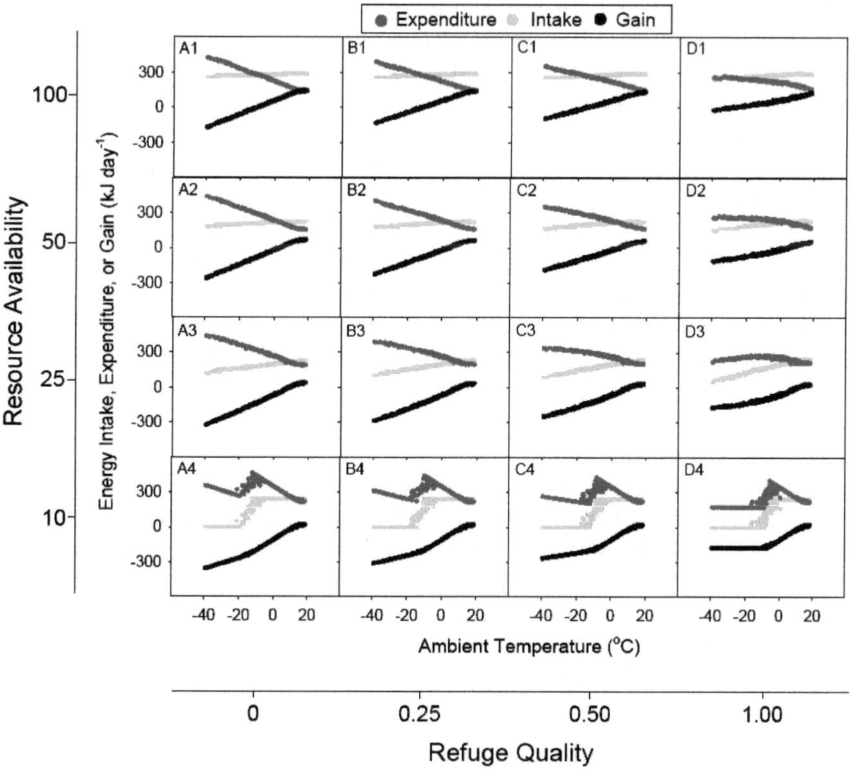

Figure 4.3. The effect of resource availability (R) and refuge quality (Q) on the relationship between ambient temperature (T_a) and daily intake (I), expenditure (E), and gain ($G = I - E$). See Figure 4.2 for explanation of parameters, resource, and refuge scenarios.

range of T_a. When resources are very abundant, I varies little with T_a because satiation occurs quickly at all temperatures. At intermediate resource levels, I tends to increase with T_a either linearly or according to diminishing returns, because although activity increases curvilinearly with T_a, intake per unit time spent active is characterized by diminishing returns due to the satiety imposed by Eq. (4.17).

The expected relationship between energy expenditure (E) and T_a varies widely according to refuge quality and resource availability (Figure 4.3 dark gray symbols). In general, E should decrease with increasing T_a due to reduced thermoregulatory requirements. However, because activity increases expenditure due to both activity costs and higher thermoregulatory requirements outside the refuge, the negative relationship between E and T_a can be weakened or even reversed by the tendency for activity to increase with T_a (Figure 4.2). This reversal is strongest under a combination of moderately low resource availability and high refuge quality, where gradual but pronounced increases in activity with T_a generate a small but consistent increase in E with T_a up to about 0°C (Figure 4.3D3). This prediction matches the empirical E vs. T_a relationship and resource and refuge characteristics of red squirrels in the wild (Humphries et al. 2005). When resources are very low and animals are constantly inside the refuge at low T_a and constantly outside the refuge at high T_a, E increases rapidly with T_a between the no activity and maximum activity thresholds (Figure 4.3A4–D4). In other situations where the more typical negative relationship between E and T_a prevails, the slope of the relationship flattens as refuge quality increases (For e.g. Figure 4.3A1–D1).

Despite the resource-, refuge-, and activity-mediated complexities in how I and E vary with T_a, expected net energy gain (G; where $G = I - E$) invariably increases with T_a (Figure 4.3, black symbols). Across all refuge and resource scenarios considered, this positive G vs. T_a relationship is approximately linear, with refuge quality affecting the slope and resource availability affecting the elevation. The only notable exception to this linearity is when low resources, a high-quality refuge, and no activity, combine to generate a negative gain rate that is independent of T_a below the T_a activity threshold (Figure 4.3D4).

Energetic efficiency (G/E) increases consistently and curvilinearly with T_a (Figure 4.4), except when resources are low and T_a is below activity thresholds (Figure 4.4A4–D4). Resource availability influences the intercept of these curves, with high resources resulting in high efficiencies. Refuge quality influences the slope of these curves, with high-quality refuges resulting in shallower slopes.

4.4.2.3 Daily, monthly, and seasonal variation in activity and energetics

The relationships between daily temperature and the various activity and energetic measures manifest into unique temporal patterns of variation depending on resource availability and refuge quality, as well as the temporal scale considered. Daily activity is always higher and usually less responsive to daily and monthly fluctuations in temperature during warm, summer months than cold, winter months (Figure 4.5). The extent of the seasonal difference is little affected by refuge quality

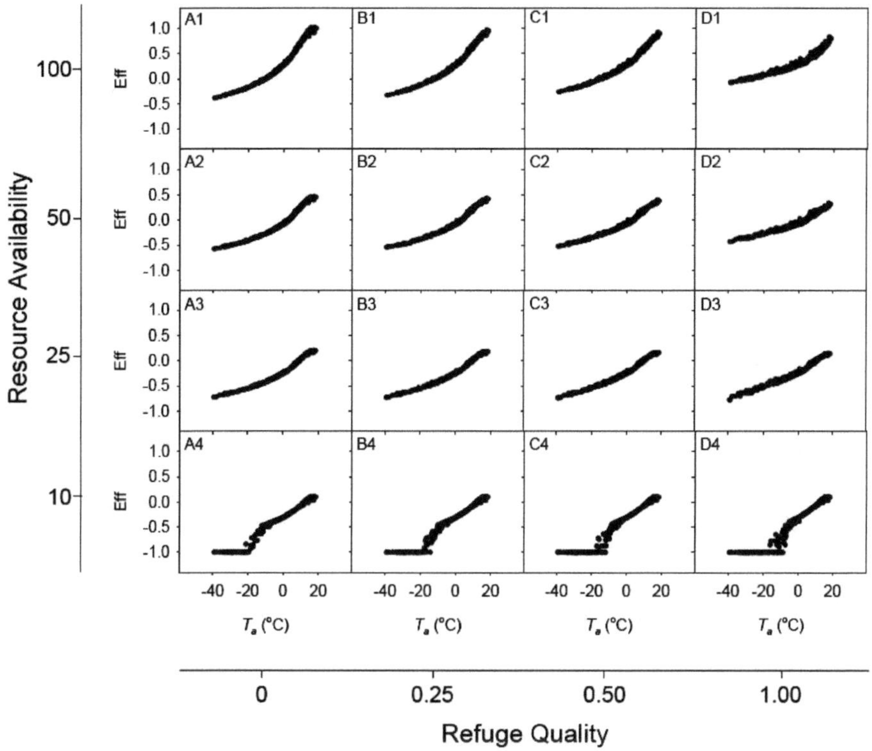

Figure 4.4. The effect of resource availability (R) and refuge quality (Q) on the relationship between ambient temperature (T_a) and daily efficiency ($(I - E)/E$). See Figure 4.2 for explanation of parameters, resource, and refuge scenarios.

but strongly affected by resource availability; high resources minimize the seasonal difference by elevating winter activity and reducing summer activity relative to activity levels observed under low resources. Daily variation in activity is most extreme during winter under conditions of low resource availability, with animals tending towards maximum activity on warm winter days and no activity on cold winter days. Energy intake is similar to seasonal activity patterns, in that intake is higher during warm summer months than cool winter months and its daily and seasonal variation is little affected by refuge quality but strongly affected by resource availability (Figure 4.6). Seasonal patterns in energy expenditure vary according to resources (Figure 4.7). When resources are high, energy expenditure mirrors temperature variation, with low expenditure during warm, summer months and high expenditure during cool, winter months (Figure 4.7A–B). However, when resources are low, energy expenditure is lowest during summer months (due to low thermoregulatory requirements) and the coldest winter months (due to low activity), and highest during spring and autumn (due to higher thermoregulatory costs than

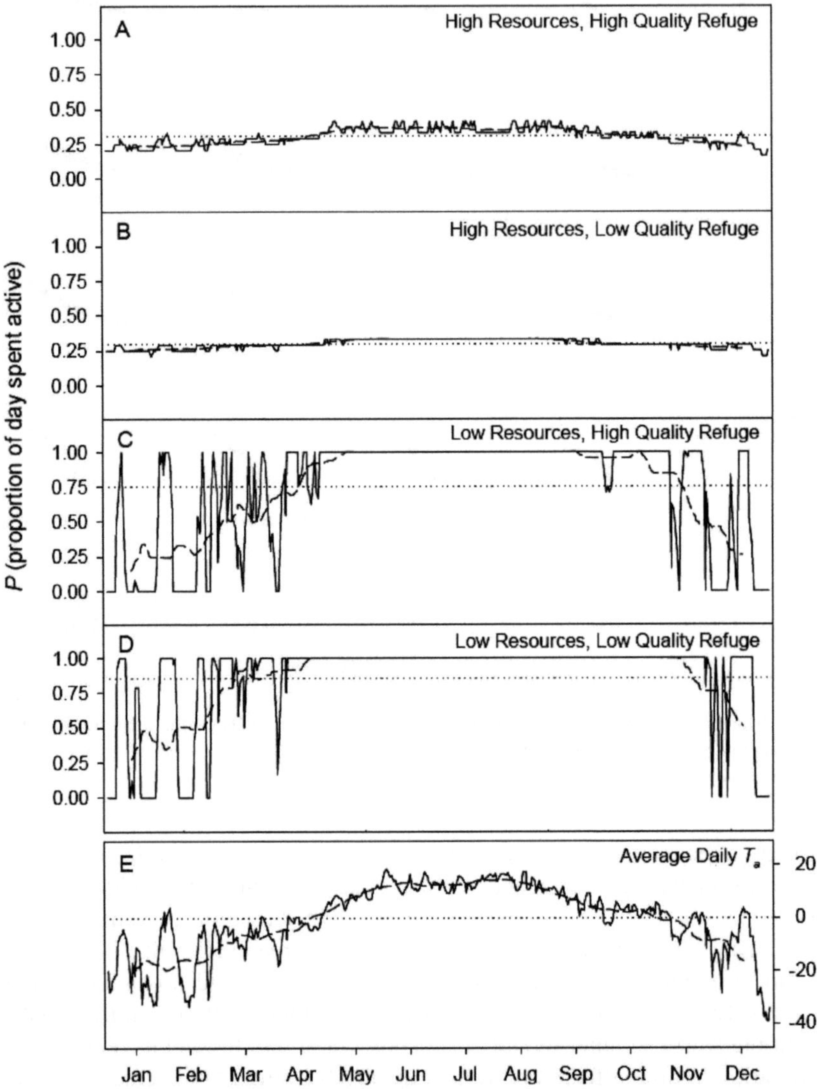

Figure 4.5. Time series of ambient temperature (T_a) and activity (p; proportion of 24 h spent active) for four resource and refuge scenarios. (a) High resources ($R = 100, \alpha = 1.50$) and high-quality refuge ($Q = 1$). (b) High resources ($R = 100, \alpha = 1.50$) and low-quality refuge ($Q = 0$). (c) Low resources ($R = 10, \alpha = 1.0$) and high-quality refuge ($Q = 1$). (d) Low resources ($R = 10, \alpha = 1.0$) and low-quality refuge ($Q = 0$). (e). Average daily air temperature (°C).

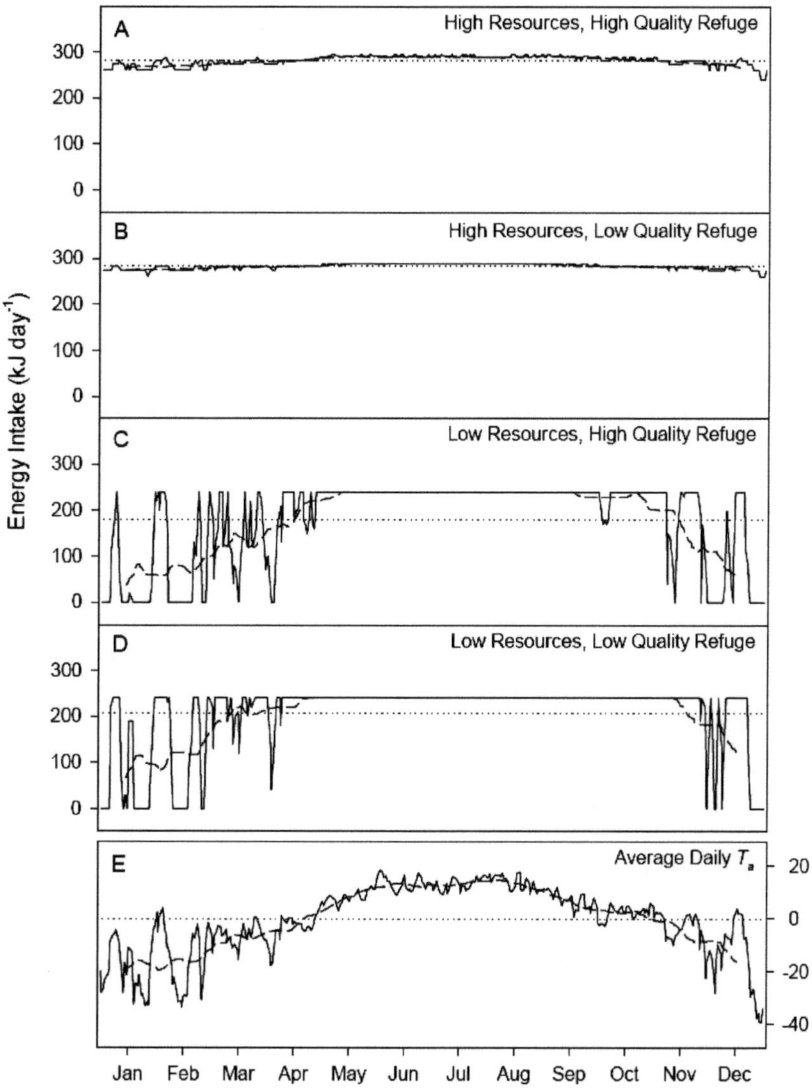

Figure 4.6. Time series of ambient temperature (T_a) and intake (I; kJ/day) for four resource and refuge scenarios (see Figure 4.4).

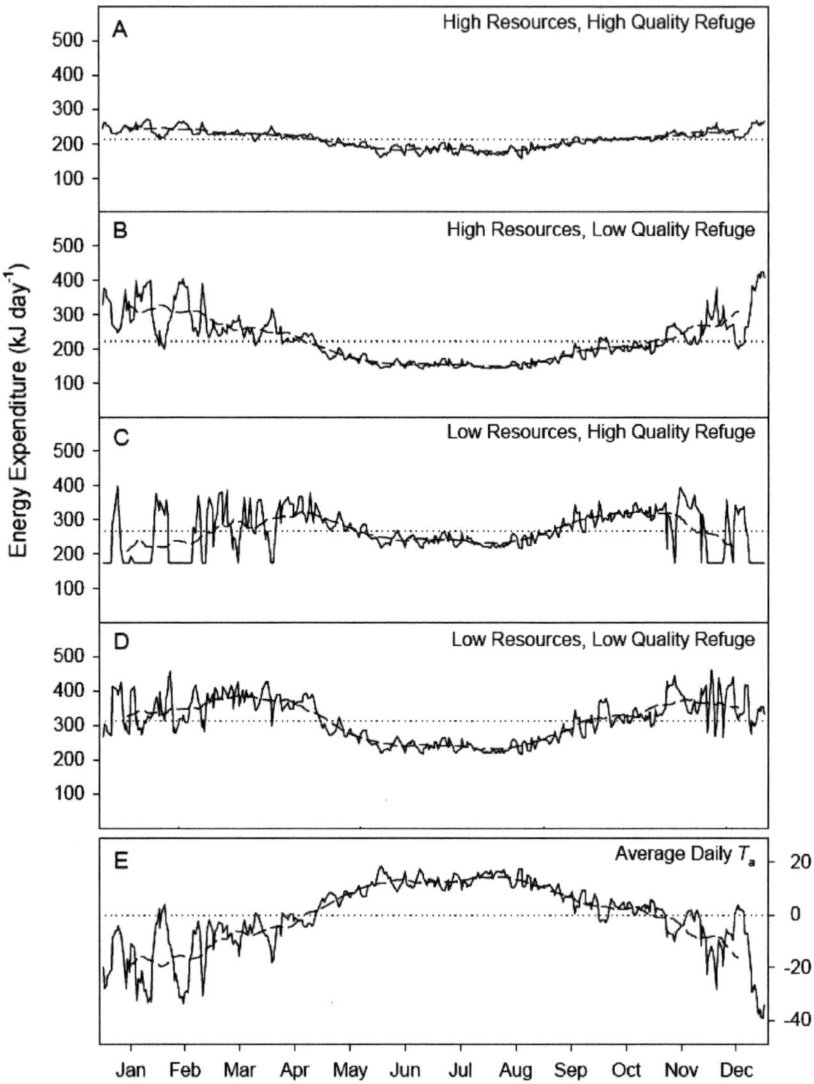

Figure 4.7. Time series of ambient temperature (T_a) and expenditure (E; kJ/day) for four resource and refuge scenarios (see Figure 4.4).

in summer and higher activity costs than in winter; Figure 4.7C–D). Temporal variation in energy gain is the most consistent and intuitive (Figure 4.8). Gain is always higher during warm summer months than cool winter months. Resource availability primarily affects average gain rates and refuge quality determines the

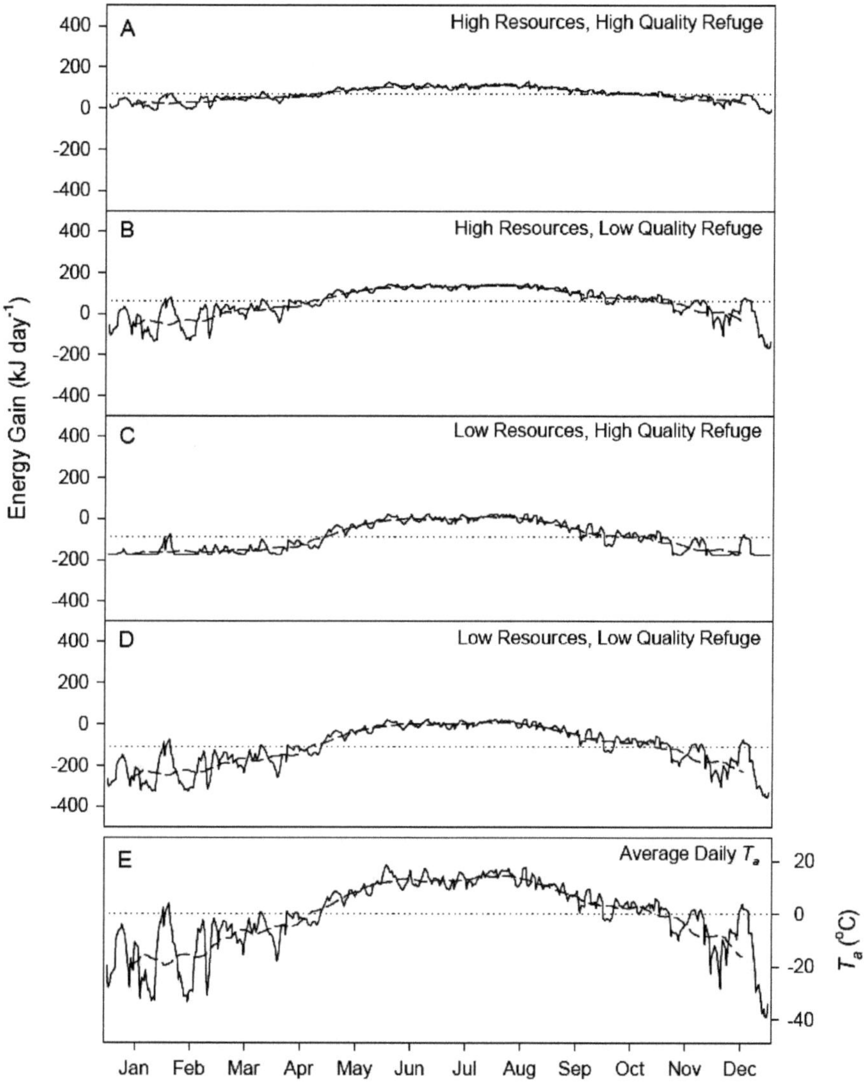

Figure 4.8. Time series of ambient temperature (T_a) and gain ($G = I - E$; kJ/day) for four resource and refuge scenarios (see Figure 4.4).

extent of daily and seasonal fluctuations in these gains. Efficiency follows the same seasonal pattern as gain rates, but is characterized by much more short-term variability, particularly when resources are abundant and refuge quality is high (Figure 4.9).

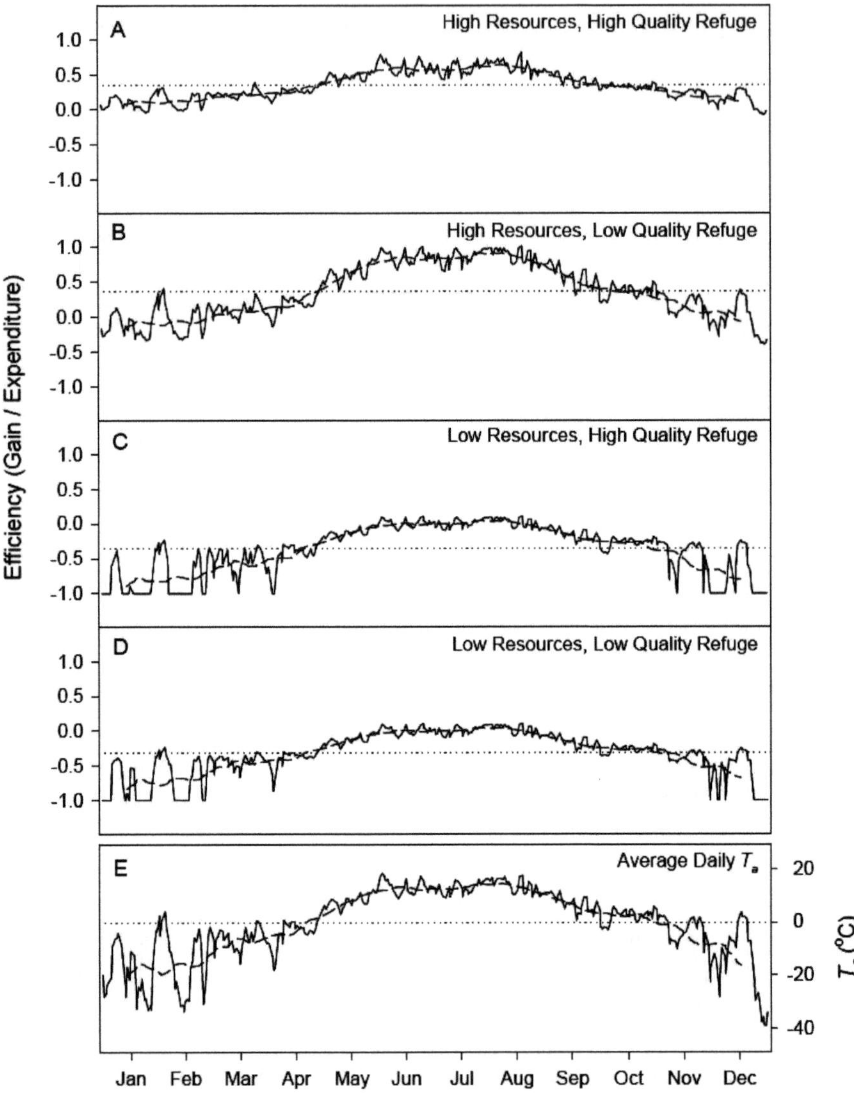

Figure 4.9. Time series of ambient temperature (T_a) and efficiency ($(I - E)/E$) for four resource and refuge scenarios (see Figure 4.4).

4.4.2.4 *Spectral comparison of activity, energetics, and ambient temperature*

The spectra of endotherm activity and energetic traits responding optimally to a reddened T_a spectrum can become substantially whitened if animals have access to high, satiating resources and high-quality, thermal refuges (Figure 4.10). In general, the colour spectrum of gain is most similar to T_a, because gain is the currency that

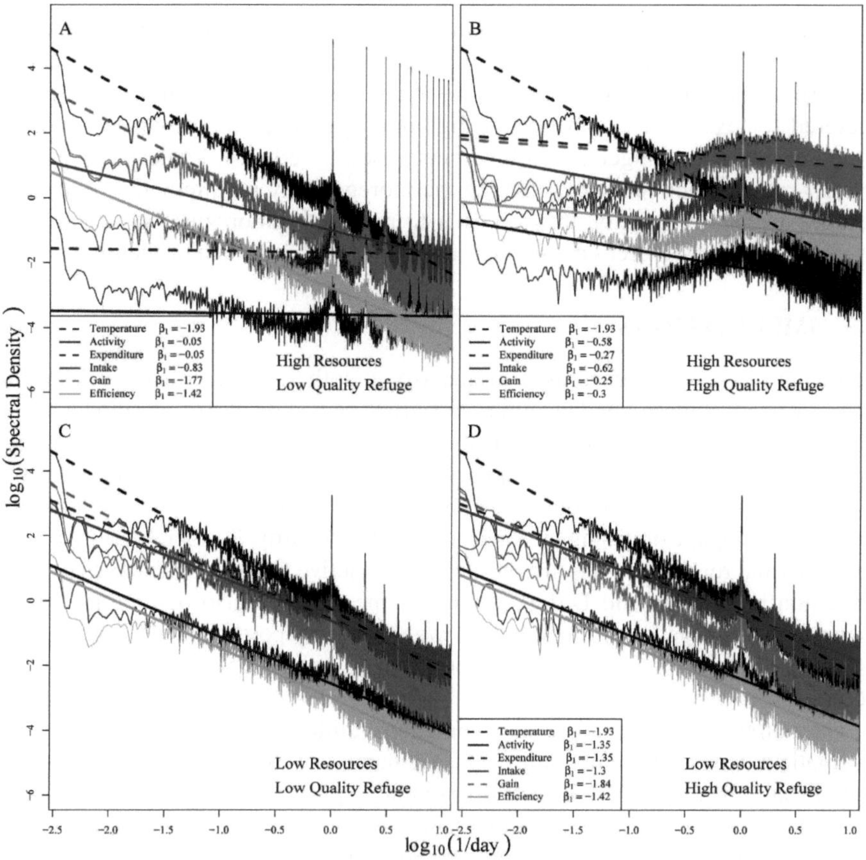

Figure 4.10. Spectral densities of activity and energetic traits in comparison to ambient temperature (T_a) for four resource and refuge scenarios. See Figures 4.1–4.10 for explanation of traits and resource/refuge scenarios.

is being optimized and there are strong thermal constraints on this optimization. In contrast, non-optimized traits such as activity, expenditure, and efficiency are more likely to decouple from T_a as animals differentially adjust these traits to optimize gain under different environmental circumstances. When resources are abundant and refuge quality is low, activity, expenditure, and efficiency have whitened spectra (Figure 4.10A), because daily saturation of intake compresses their multi-day fluctuations by deterring long periods of activity on warm days and promoting short periods of activity on cold days. When abundant resources coincide with high-quality refuges, all traits have whitened spectra (Figure 4.10B), because, in addition to the long-term filtering effect described above, all traits amplify short-term variation in T_a due to the pronounced dichotomy in the energetics of inactive and active animals triggered by small changes in T_a. High resources and high-quality refuges are thus both necessary for gain to effectively decouple from T_a fluctuations, because abundant resources enable the rapid resource acquisition and satiation required for extended occupation of high-quality refuges. If resources are scarce, foraging times are longer and refuge occupancy is reduced, rendering even high-quality refuges ineffective at reducing thermoregulatory requirements. Whereas, if refuges are of low quality, they do little to reduce thermoregulatory requirements even if abundant resources cause them to be occupied for much of the time.

4.5 IMPLICATIONS

Although many approaches used to quantify endotherm metabolism seek to minimize variation imposed by behaviour, the structure and results of our model suggest behaviour plays a central role in shaping the energetic response of endotherms to environmental variability. The simple behavioural choice between being inactive in a thermal refuge or active out of a refuge is the primary determinant of whether endotherms engage or disengage environmental variability. Further, because occupying a thermal refuge also requires suspension of active foraging and serves to reduce predation risk (Lima and Dill 1990), this same decision also determines whether endotherms engage or disengage food webs. The proportion of time spent active is thus a variable of fundamental importance to animal energetics and metabolic ecology.

The output of our model reveals something quite paradoxical about endotherm energetic responses to temperature. On one hand, endotherms have many energetic options at their disposal, and their differential exploitation of these options, across the wide range of environmental temperatures encountered, generates highly variable and counter-intuitive energetic responses. For example, if resources and refuge quality are low, energy expenditure is predicted to decrease, then increase, then decrease again as T_a increases (Figure 4.3A4–B4), resulting in reduced energy expenditure during the warmest summer months and the coldest winter months and elevated energy expenditure during spring and autumn (Figure 4.5C–D). On the other hand, in contrast to the wide variety of thermal impacts on their activity, intake, and expenditure, the energy gain and efficiency of endotherms invariably increases with rising air temperature, regardless of resource availability and refuge quality. Thus, despite activity optimization and access to thermal refuges, homeothermic endotherms are remarkably unsuccessful in avoiding the simple biological constraints that make life more difficult in the cold.

The consistency of the relationship between T_a and net energy gain provides little reason for population ecologists, who are primarily interested in the biomass accumulation of consumer populations, to concern themselves with the details of thermal refuges, activity optimization, and the various terms within an animal's energy budget. We will not attempt to convince them otherwise here. In fact, we will point out that they could take Eq. (4.13), assume that I will not vary with T_a (since we held resources constant throughout the year) and that E will vary linearly with T_a according to resting thermal conductance, to predict

$$G \alpha C T_a. \tag{4.18}$$

This simple equation comes remarkably close to replicating our much more complex model's predictions. For the parameter values listed in Figure 4.3, Eq. (4.18) predicts $G\alpha 4.6T_a$, while the least-squares linear equation for the composite of the 16 gain curves in Figure 4.3 is:

$$G = -26.309 + 5.12T_a \tag{4.19}$$

Thus, consistent with the general modelling philosophy of metabolic ecology, adequate ecological predictions can be obtained with simple, mechanistic models based on body size and temperature alone. As is frequently the case, adequate prediction of *what* will happen does not necessarily require detailed knowledge of *how* it will happen (Peters 1991; Pigliucci 2003).

However, from a community/trophic ecology perspective, details about how energetic gains are realized in variable environments become much more interesting than the gains themselves. Although the growth of a consumer population depends only on its net gain $(I - E)$, its facilitation of energy transfer from resources to predators depends primarily on its net efficiency (G/E) and activity, both of which are strongly influenced by resource availability and refuge quality. Efficiency is important because this dictates the amount of consumer biomass generated per resource biomass expended. Activity is important because it is a primary determinant of efficiency and, when thermal refuges also serve as predator refuges, how effectively energy transfers from consumers to predators.

Our model is based on the assumption that endotherms seek to maximize net gain rather than energetic efficiency. Thus, although efficiency may not concern an optimizing endotherm, it is, or at least should be, a major concern to this endotherm's resources and predators. The evolution of endothermy provides an excellent macroevolutionary example of selection for high gains over high efficiencies. Detailed studies of the annual energetics of golden-mantled ground squirrels by Kenagy et al. (1989) provide an example of the net annual gains and efficiencies of a typical small endotherm (Table 4.1). Annual energy gains are trivial relative to the sizeable energy intake and energy expenditure required to obtain these gains. This should not be surprising, given that endotherms are determinate growers characterized by relative small annual reproductive output. There is, literally, nowhere else for all that ingested energy to go. But the values in Table 4.1 serve as a reminder that when Yodzis and Innes (1992) state that an animal "acquires energy from its environment ... with the end result that a portion of the ingested energy is converted into biomass," that this end result is only a tiny fraction of the ingestion and metabolism generating the gain.

In addition to I and E being much bigger than G, they also are likely to be much more variable, because selection should cause many animals to accept large changes in I and E for very small changes in G. For example, eastern chipmunks (*Tamias striatus*) respond to supplemental food by substantially reducing torpor expression

Table 4.1. Annual energetic parameters for free-ranging golden-mantled ground squirrels (calculated from Kenagy et al. 1989)

Age/sex	Intake (kJ)	Expenditure (kJ)	Gain (kJ)	Efficiency
Yearling males	38,536	38,434	102 (100% growth)	0.0026
Adult males	42,064	42,054	10 (100% gametes)	0.00024
Yearling females	44,982	42,597	2,385 (5% growth, 4% fetuses, 91% milk)	0.056
Adult females	44,238	41,943	2,295 (1% growth, 4% fetuses, 95% milk)	0.055

during hibernation, which leads to five- to tenfold increases in their overwinter E and therefore I (Munro et al. 2005). We have speculated elsewhere that these highly consumptive energetic responses to surplus resources may be adaptive due to physiological and ecological costs of torpor that have subtle effects on lifetime reproductive success (Humphries et al. 2003). A five- to tenfold increase in resource consumption that provides a small, probabilistic increase in individual fitness may be evolutionary adaptive, but is highly inefficient from a trophic perspective. Presumably the predators that prey on chipmunks would prefer there to be many more, slightly less optimized chipmunks in their home range.

Overall, our analyses suggest increases in T_a cause endotherms to simultaneously increase their net energy gain as well as the efficiency of this gain. Thus, warming T_a increases both the amount of energy endotherms accumulate as well as the efficiency of this accumulation, both of which contribute to effective energy transfer across trophic levels. This is exactly the opposite of the situation in ectotherms, where warming T_a tends to reduce energy gain and efficiency (Yodzis and Innes 1992; Vasseur and McCann 2005). This warming benefit for endotherms and warming detriment for ectotherms is somewhat paradoxical, given that the ratio of endotherm: ectotherm biomass tends to be higher in cold climates than in warm climates.

The solution to the paradoxical predominance of endotherms in cold climates may involve their enhanced capacity to energetically filter long-term fluctuations in T_a. An unusual attribute of climate is that its variance generally increases as its mean decreases. For example, daily and seasonal T_a variability increases with latitude while average annual T_a declines. As a result, adapting to cold climates requires also adapting to temporally variable climates. This temporal variability becomes particularly problematic when the period of its fluctuations begin to exceed the thermal inertia of individuals or populations (May 1976; Vasseur, Chapter 3). The thermal inertia of individuals will be defined by their physiological and behavioural capacity to buffer body temperature and somatic condition against temperature fluctuations, whereas the thermal inertia of populations will be defined by their ecological capacity to buffer demographics against temperature fluctuations. The latter will involve longer temporal scales than the former, but thresholds for both will be extended by adaptations such as endogenous heat production, expanded aerobic capacity in combination with energy storage, and large body size. Of course, it is precisely these traits that differentiate endotherms from ectotherms and high-latitude endotherms from low-latitude endotherms (e.g. Bergmann's rule, Meiri and Dayan 2003; latitudinal gradients in resting metabolism, Lovegrove 2003). We have shown here that for small, high-latitude endotherms, thermal refuges and activity optimization are additional, essential contributors to their capacity to amplify short-term temperature variability and filter long-term temperature variability. Small endotherms that hibernate will have additional capacities, beyond those included in our model, to filter long-term temperature fluctuations.

Thus, high-latitude endotherms can be viewed as world-class specialists in filtering long-term temperature fluctuations that would otherwise compromise individual homeostasis or population growth. Achieving this long-term buffering capacity requires energetic and behavioural traits that remain highly influenced by temperature and an overriding, energetically inefficient design that is particularly inefficient in cold climates. Accordingly, we hypothesize that endothermy is detrimental for cold average conditions but advantageous for the temporal variability that co-occur with these conditions. In relation to this hypothesis, it is notable that endotherms evolved and radiated largely outside of the world's oceans and large water bodies, which are environments where thermal conductance is enhanced, thermal refugia are unavailable, and cool average temperatures do not co-occur with temporally variable temperatures.

4.6 ACKNOWLEDGEMENTS

Murray M. Humphries thanks Peter Yodzis for providing some bioenergetic toeholds in the cliff of theoretical ecology, and acknowledges the financial support of ArcticNet and NSERC's Northern Research Chairs and Discovery Grants Programs. James Umbanhowar acknowledges the support of NSERC grants to Kevin McCann.

4.7 LITERATURE CITED

Anderson, K.J. and W. Jetz. 2005. The broad-scale ecology of energy expenditure of endotherms. *Ecology Letters* 8(7): 310–318.

Bakken, G.S. 1976. A heat transfer analysis of animals: unifying concepts and the application of metabolism chamber data to field ecology. *Journal of Theoretical Biology* 60: 337–384.

Blaxter, K. 1989. *Energy metabolism in animals and man.* Cambridge University Press, Cambridge.

Brown, J.H., J.F. Gillooly, A.P. Allen, V.M. Savage, and G.B. West. 2004. Towards a metabolic theory of ecology. *Ecology* 85: 1771–1789.

Chappell, M.A. and G.A. Bartholomew. 1981. Standard operative temperatures and thermal energetics of the antelope ground squirrel *Ammospermophilus leucurus*. *Physiological Zoology* 54: 81–93.

Cushing, J.M., B. Dennis, R.A. Desharnais, and R.F. Costantino. 1998. Moving toward an unstable equilibrium: saddle nodes in population systems. *Journal of Animal Ecology* 67: 298–306.

De Roos, A.M. and L. Persson. 2001. Physiologically structured models – from versatile technique to ecological theory. *Oikos* 94: 51–71.

Dzialowski, E.M. 2005. Use of operative temperature and standard operative temperature models in thermal biology. *Journal of Thermal Biology* 30: 317–334.

Gillooly, J.F., J.H. Brown, G.B. West, V.M. Savage, and E.L. Charnov. 2001. Effects of size and temperature on metabolic rate. *Science* 293: 2248–2251.

Glazier, D.S. 2005. Beyond the '3/4-power law': variation in the intra- and interspecific scaling of metabolic rate in animals. *Biological Reviews* 80: 611–662.

Greenman, J.V. and T.G. Benton. 2003. The amplification of environmental noise in population models: causes and consequences. *American Naturalist* 161: 225–239.

Guppy, M. and P. Withers. 1999. Metabolic depression in animals: physiological perspectives and biochemical generalizations. *Biological Reviews* 74: 1–40.

Halley, J.M. 1996. Ecology, evolution and $1f$-noise. *Trends in Ecology and Evolution* 11: 33–37.

Huey, R.B. 1991. Physiological consequences of habitat selection. *American Naturalist* 137: S91–S115.

Humphries, M.M. and S. Boutin. 2000. The determinants of optimal litter size in free-ranging red squirrels. *Ecology* 81: 2867–2877.

Humphries. M.M., S. Boutin, D.W. Thomas, J.D. Ryan, C. Selman, A.G. McAdam, D. Berteaux, and J.R. Speakman. 2005. Expenditure freeze: the metabolic response of small mammals to cold environments. *Ecology Letters* 8: 1326–1333.

Humphries, M.M., D.W. Thomas, and D.L. Kramer. 2003. The role of energy availability in mammalian hibernation: a cost-benefit approach. *Physiological and Biochemical Zoology* 76: 165–179.

Kam, M. and A.A. Degen. 1997. Energy requirements and the efficiency of utilization of metabolizable energy in free-living animals: evaluation of existing theories and generation of a new model. *Journal of Theoretical Biology* 184: 101–104.

Kenagy, G.J., S.M. Sharbaugh, K.A. Nagy. 1989. Annual cycle of energy and time expenditure in a golden-mantled ground squirrel population. *Oecologia* 78: 269–282.

Kooijman, S.A.L.M. 2000. *Dynamic energy and mass budgets in biological systems*, 2nd ed. Cambridge University Press, Cambridge.

Lima, S.L. and L.M. Dill. 1990. Behavioural decisions made under the risk of predation: a review and prospectus. *Canadian Journal of Zoology* 68: 619–640.

Lovegrove, B.G. 2003. The influence of climate on the basal metabolic rate of small mammals: a slow fast continuum. *Journal of Comparative Physiology B* 173: 87–112.

Lovegrove, B.G. 2005. Seasonal thermoregulatory responses in mammals. *Journal of Comparative Physiology B* 175: 231–247.

McNab, B.K. 2002. *The physiological ecology of vertebrates: a view from energetics.* Cornell University Press, Cornell, NY.

McPhaden, M.J. 1999. Genesis and evolution of the 1997–98 El Nino. *Science* 283: 950–954.

May, R.M. 1976. *Theoretical ecology, principles and applications.* Blackwell Scientific, Oxford.

Meiri, S. and T. Dayan. 2003. On the validity of Bergmann's rule. *Journal of Biogeography* 30: 331–351.

Munro, D., D.W. Thomas, and M.M. Humphries. 2005. Torpor patterns of hibernating eastern chipmunks (*Tamias striatus*) vary in response to the size and fatty acid composition of food hoards. *Journal of Animal Ecology* 74: 692–700.

Pauls, R.W. 1978a. Energetics of the red squirrel: a laboratory study of the effects of temperature, seasonal acclimitization, use of the nest and exercise. *Journal of Thermal. Biology* 6: 79–86.

Pauls, R.W. 1978b. Behavioural strategies relevant to the energy economy of the red squirrel (*Tamiasciurus hudsonicus*). *Canadian Journal of Zoology* 56: 1519–1525.

Odum, E. 1963. *Ecology.* Holt, Rinehart and Winston, New York.

Peters, R.H. 1991. *A critique for ecology.* Cambridge University Press, Cambridge.

Pigliucci, M. 2003. From molecules to phenotypes? The promise and limits of integrative biology. *Basic and Applied Ecology* 4: 297–306.

Priestley, M.B. 1981. *Spectral analysis and time series.* Academic Press, London.

Robbins, C.T. 1993. *Wildlife feeding and nutrition.* Academic Press, San Diego, CA.

Scholander, P.F., R. Hock, V. Walters, and L. Irving. 1950. Adaptation to cold in arctic and tropical mammals and birds in relation to body temperature, insulation, and basal metabolic rate. *Biological Bulletin* 99: 225–236.

Sibly, R.M. and P. Calow. 1986. *Physiological ecology: an evolutionary approach.* Blackwell, Oxford.

Spicer, J.I. and K.J. Gaston. 1999. *Physiological diversity and its ecological implications.* Blackwell Scientific, Oxford.

Speakman, J.R. 2000. The cost of living: field metabolic rates of small mammals. *Advances in Ecological Research* 30: 177–297.

Stenseth, N.C., A. Mysterud, G. Ottersen, J.W. Hurrell, K.S. Chan, and M. Lima. 2002. Ecological effects of climate fluctuations. *Science* 297: 1292–1296.

Rind, D. 2002. The sun's role in climate variations. *Science* 296: 673–677.

van der Meer, K. 2006. Metabolic theories in ecology. *Trends in Ecology and Evolution* 21: 136–140.

Vasseur, D.A. and K.S. McCann. 2005. A mechanistic approach for modeling temperature-dependent consumer-resource dynamics. *American Naturalist* 166: 184–198.

Vasseur, D.A. and P. Yodzis. 2004. The color of environmental noise. *Ecology* 85: 1146–1152.

Walsberg, G.E. 1985. Physiological consequences of microhabitat selection. In: *Habitat selection in birds* (ed., M.L. Cody), pp. 389–413. Academic Press, Orlando, FL.

Ydenberg, R.C., C.V.J. Welham, R. Schmidhempel, P. Schmidhempel, and G. Beauchamp. 1994. Time and energy constraints and the relationship between currencies in foraging theory. *Behavioral Ecology* 5: 28–34.

Yodzis, P. and S. Innes. 1992. Body size and consumer-resource dynamics. *American Naturalist* 139: 1151–1175.

CHAPTER 5

ENVIRONMENT FORCING POPULATIONS

ESA RANTA*, VEIJO KAITALA, AND MIKE S. FOWLER

Integrative Ecology Unit, Department of Biological and Environmental Sciences, P.O. Box 65 (Viikinkaari 1), FIN-00014 University of Helsinki, Finland
**E-mail: esa.ranta@helsinki.fi, Fax: + 358 9 191 57694*

AND

JAN LINDSTRÖM

Graham Kerr Building, Division of Environmental and Evolutionary Biology, Institute of Biomedical and Life Sciences, University of Glasgow, Glasgow, G12 8QQ, Scotland

5.1 ABSTRACT

A striking feature for many species is that populations fluctuate in synchrony over vast geographical ranges. Historically synchronicity was attributed to external forcing (like weather) that influences local population renewal. It is now understood that forcing can be due to any agent (dispersal, competition, predator–prey interactions) interfering with local processes. Data show that often synchrony levels off with increasing distance among populations. Such a feature suggests presence of population travelling waves. That is, processes of population highs coinciding in

David A. Vasseur and Kevin S. McCann, The Impact of Environmental Variability on Ecological Systems, 89–110.

time in limited regions while elsewhere in space population lows coincide and the pattern changes in location over time. A poorly studied feature in this context is synchrony across species that is suggested to emerge due to interspecific interaction networks. That external forcing causes populations to fluctuate in step is easy to prove. However, time series of population change do not always necessarily carry any extractable information about the agent that is forcing the population, unless complemented with auxiliary data about the environment and the focal population studied. The fingerprint of the external signal is more easily visible in the noise-modulated population data if the underlying renewal process is undercompensatory.

Keywords: synchrony, Moran effect, population renewal, density dependence, external forcing, travelling wave.

5.2 INTRODUCTION

Understanding causes and consequences of population fluctuations is the major concern of contemporary population ecology. The debate has been intensive on the relative merits of density-dependent and density-independent population regulation (Lack 1954; Andrewartha and Birch 1954; Murray 1982; Turchin 2003; Cappuccino 2005). Interest in the way climate (or weather) may influence population dynamics dates back to climate control theory (Bodenheimer 1938). Motivated in particular by research on insect populations and game animal dynamics, climate control theory advocates that population fluctuations of many species can be explained by weather/climate factors (Stenseth et al. 1999; Sæther et al. 2000; Holmgren et al. 2001; Jaksic 2001; Ottersen et al. 2001; Post et al. 2001; Blenckner and Hillebrand 2002; Jonzén et al. 2002b; Barbraud and Weimerskirch 2003). Perhaps the best-known example is the proposed causal link between sunspot cycle-derived weather conditions and the dynamics of snowshoe hare and Canada lynx (Elton 1924; Sinclair et al. 1993; but see Ranta et al. 1997a for an orthogonal view).

Research into how weather (as climatic variation, or more simply, some external perturbation) influences local populations can be split into two main subject areas:

- In the more straightforward tradition the rationale is to seek for weather and climate variables that putatively correlate with long-term data on population dynamics. This is likely to arise from the temptation to find biologically rational explanations for the effect of almost any weather-derived variable on population fluctuations via mortality and reproduction, the key elements of population persistence. The origin of the idea is in the climatic control theory (Bodenheimer 1938; Andrewartha and Birch 1954), stating that populations are strongly influenced by weather, and are thus regulated by these factors.
- The second approach, population synchrony research, focuses on consequences of external perturbation on the degree of the temporal match in population

fluctuations over vast geographic ranges. In this approach there is no assumption of whether climate regulates populations. Rather, the focus is on the significance of space-time dimension in influencing population patterns. This approach seeks answers to the following questions: Why do spatially separated populations fluctuate in synchrony? Why does the degree of synchrony often decrease with increasing distance between populations? For what reason, and how does the synchrony vary in time? In explaining synchrony patterns, this tradition does not rely on external perturbations only, but brings in another synchronizing factor, dispersal between subpopulations or localities. In fact, we shall end up arguing that any mechanism that disturbs local density-dependent feedback in the population renewal process is likely to cause synchrony given it is influencing more than one local population.

We shall discuss here, using both theory and empirical examples, the issues that we consider of relevance in the interplay of single-species populations and environmental fluctuations, with data replicated over time and space.

5.2.1 Evolution of environmental effects in ecology

Haeckel (1870) was keen on promoting a branch of natural history, which he referred to as "the total relations of the animal both to its inorganic and organic environment," later known as the science of ecology. Andrewartha and Birch (1954), two key players in the debate on the relative significance of density-dependent and density-independent factors in influencing population fluctuations, made an effort to clarify what the "environment" in Haeckel's definition really meant. Their aim was to find a meaning that would be of use in understanding and explaining the observed distributions and abundances of organisms in nature. For pragmatic reasons, they considered that the "environment" should be broken into components that can be treated separately and that render their study possible by observation and experiment. With these operative constraints in mind, they proposed the environment to be composed of four components: weather, food, other organisms (including disease-causing organisms), and a place to live.

We cannot undervalue the significance of weather (or climate in more general terms) in influencing fluctuations in population numbers of any given species over its distribution range. To address this question we shall employ a conceptual model by Andrewartha and Birch (1954), to clarify our starting point in this chapter. Consider the distribution range (say from south to north) of a focal species, which is limited by adverse weather conditions in its ability to sustain viable populations. The geographical margin of the distribution range will shift over time as local conditions improve or deteriorate with environmental fluctuations (Figure 5.1a). Moving from the distribution margins towards the centre, local population size tends to increase. We may now ask; how do the dynamics of focal populations look, when taken from different regions of the distribution range?

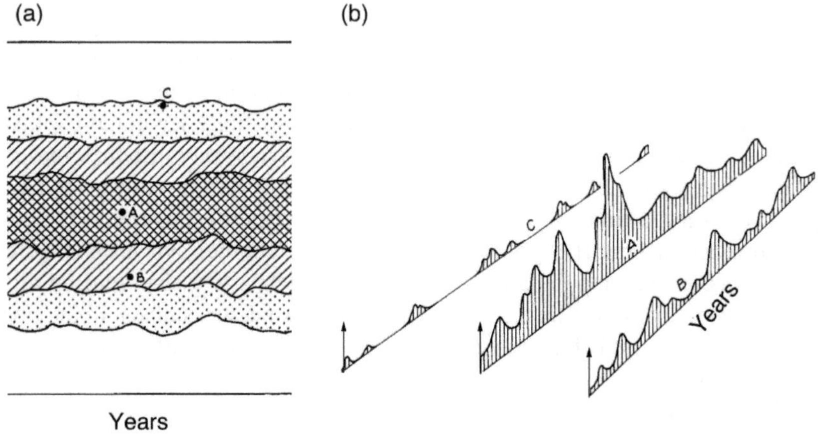

Figure 5.1. (a) Distribution range (y-axis: south–north) of a hypothetical species over time (x-axis). Average local population density is indicated by intensity of the shading. Temporal dynamics of three local populations (A, B, and C) are shown in (b). The arrow indicates direction of increasing population size, y-axis. The graph is modified after (Andrewartha and Birch 1954; their Figures 1.1 and 1.2).

5.3 CLIMATE FORCING POPULATIONS

It is perhaps understandable that a population (marked C in Figure 5.1) close to the edge of the distribution range is seldom large enough for long-term persistence. Such small populations are subject to demographic stochasticity where chance events may drive them to extinction at times, but they can be re-established by colonists in periods when conditions in the centre of the distribution range are favourable (Figure 5.1b). If so, we expect to see such a local population showing irregular dynamics that one might be able to link to particular climatic events. In contrast, populations closer to the centre of the distribution range (marked A and B in Figure 5.1) are, due to more favourable conditions, large enough to persist over time, albeit with fluctuations that are often correlated in time (Figure 5.1b). Populations A, B, and C are influenced by local processes (births and deaths) and by processes tying them into a tapestry of the larger regional network (by emigration and immigration).

In an attempt to answer our question above in a more solid way, we use a formal model. Let $X(t) = \ln[N(t)]$, where $N(t)$ refers to population size of our focal species at time t (likewise, $Y(t) = \ln[M(t)]$). For population growth between any two consecutive points in time we have $R(t) = X(t+1) - X(t)$. We assume that resources for the renewal of any population are limiting, and hence use a first-order linear process to adopt density dependence into the population renewal process:

$$R(t) = aX(t) + b. \tag{5.1}$$

Here a and b are constants. Equation (5.1), extended to two spatially distinct populations X and Y, can also be written as (Royama 1992)

$$\begin{aligned} X(t+1) &= \{[1+a_X]\ln[(1-m_X)N(t)+m_Y M(t)]+b_X\}\,\mu_X(t) \\ Y(t+1) &= \{[1+a_Y]\ln[(1-m_Y)M(t)+m_X N(t)]+b_Y\}\,\mu_Y(t) \end{aligned} \quad (5.2)$$

Here μ_X(t) and μ_Y(t) refer to external perturbations (such as climate forcing) influencing the renewal process in each population. The term m refers to the fraction of individuals dispersing. Moran (1953) showed that if the perturbation processes were correlated, the populations would fluctuate in synchrony. Specifically, the condition $\text{COR}(X,Y) = \text{COR}(\mu_X,\mu_Y)$ later became known as Moran's theorem (Royama 1992, 2005) or, in more liberal terms, as the Moran effect (Ranta et al. 2006). The $\text{COR}(\mu_X,\mu_Y)$ can also be used as a measure of the spatial proximity of the two populations (Figure 5.2a). When the correlation is high, the two populations are likely to exist in more closely matching (or geographically proximal) environments than when the correlation is low.

To reiterate the validity of the Moran's theorem we simulated two identical populations ($a_X = a_Y$, $b_X = b_Y$, and $m_X = m_Y$) but allowed the environment to vary so that $\text{COR}(\mu_X,\mu_Y)$ ranged from 1 (identical environments) to 0 (totally independent environments). The system was iterated over 1,100 time steps, using the final 100 time steps to score the synchrony $\text{COR}(X,Y)$ by using cross correlation with lag zero, r_0. The procedure was replicated 1,000 times for each parameter

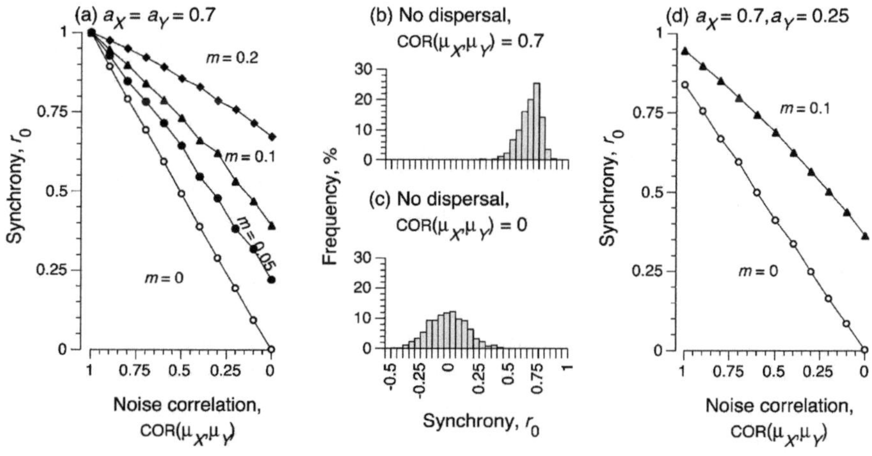

Figure 5.2. (a) Synchrony in fluctuations of two identical populations ($a_X = a_Y$, $b_X = b_Y$, and $m_X = m_Y$), X and Y, measured as sample cross correlation with lag zero, r_0, as a function of sample correlation between the environments, COR(μ_X,μ_Y) the two populations live. Open rings and connecting line indicate the case of no dispersal (m = 0), while the closed symbols refer to systems with dispersal (values of m are inserted). The data are averages over 1,000 replicated runs for each parameter combination. Frequency distributions of the sample cross correlations, r_0, for two values of environmental correlation, (b) COR(μ_X,μ_Y) = 0.7, (c) COR(μ_X,μ_Y) = 0. The averages of the two frequency distributions are 0.7 and 0, respectively (note the wide spread of the two histograms). (d) As (a) but for two populations differing somewhat in their dynamics, a_X = 0.7, a_Y = 0.25. Note that even a small amount of dispersal enhances substantially the overall level of synchrony.

combination. With $m = 0$, the sample mean r_0 equalled $\mathrm{COR}(\mu_X,\mu_Y)$. It is worth noting that there was large variation in the r_0 values among replicates, as indicated for $\mathrm{COR}(\mu_X,\mu_Y) = 0$ and $\mathrm{COR}(\mu_X,\mu_Y) = 0.7$ (Figure 5.2b,c). Importantly, the wide spread of sample synchrony estimates warns us not to draw too hasty conclusions of a lack of synchronicity if the sample mean synchrony calculated from the data is low. Also, as expected, with dispersal ($m > 0$) synchrony in population fluctuations is further enhanced. By assuming $a_X \neq a_Y$ one can emulate a situation where the two populations actually differ somewhat in their biology (for demonstrative purposes, we used a rather extreme difference, $a_X = 0.7$ and $a_Y = 0.25$) in different parts of the distribution range. As expected, Moran's theorem no longer holds under these conditions, although synchrony in population fluctuations, r_0, improves with increasing $\mathrm{COR}(\mu_X,\mu_Y)$ (see also Greenman and Benton 2001; Engen and Sæther 2005). Allowing redistribution via dispersal also enhances the synchrony level in this case (Figure 5.2d).

5.3.1 Synchrony data

Synchrony in population fluctuations on many species over their geographical ranges is well documented (Moran 1953; Ranta et al. 1995, 1997b, 1999, 2006; Cattadori and Hudson 1999; Cattadori et al. 2000; Liebhold et al. 2004). The number of known taxa displaying spatio-temporal synchrony is large and continues to increase annually (Liebhold et al. 2004). Here, we add 25 years (1982–1996) of catch statistics from five species of salmon in the Pacific North-West to this catalogue (Figure 5.3a; E. Ranta, unpublished). These data give good support (excepting perhaps Chinook salmon) for the presence of reasonably high level of synchrony in population numbers (Figure 5.3), as well as to the common finding that the overall level of synchrony often declines with increasing distance among populations (Ranta et al. 1995, 1999, 2006).

There are a few explanations for the pattern of reduced synchrony over longer distances (Paradis et al. 1999; Ranta et al. 1999, 2006; Bjørnstad and Bolker 2000). It may be a product of reduced correlation in environments as the distance between the sampled populations increases. Alternatively, decreasing synchrony with increasing distance is a product of the number of dispersing individuals declining with increasing distance. It has also been suggested that dispersal and spatially correlated environments dominate at different spatial scales (Paradis et al. 1999; but see below). That the pattern of negative correlation between synchrony and distance can also arise due to an interaction between these two mechanisms further complicates identification of the main driving factor. This can be seen from the synchrony data (Figure 5.2a) by taking $\mathrm{COR}(\mu_X,\mu_Y)$ as a proxy of distance, large values indicating closely situated populations (displaying high synchrony) and small values indicating distant populations with little synchrony. Topping off the system with the redistribution of individuals enhances the degree of synchrony.

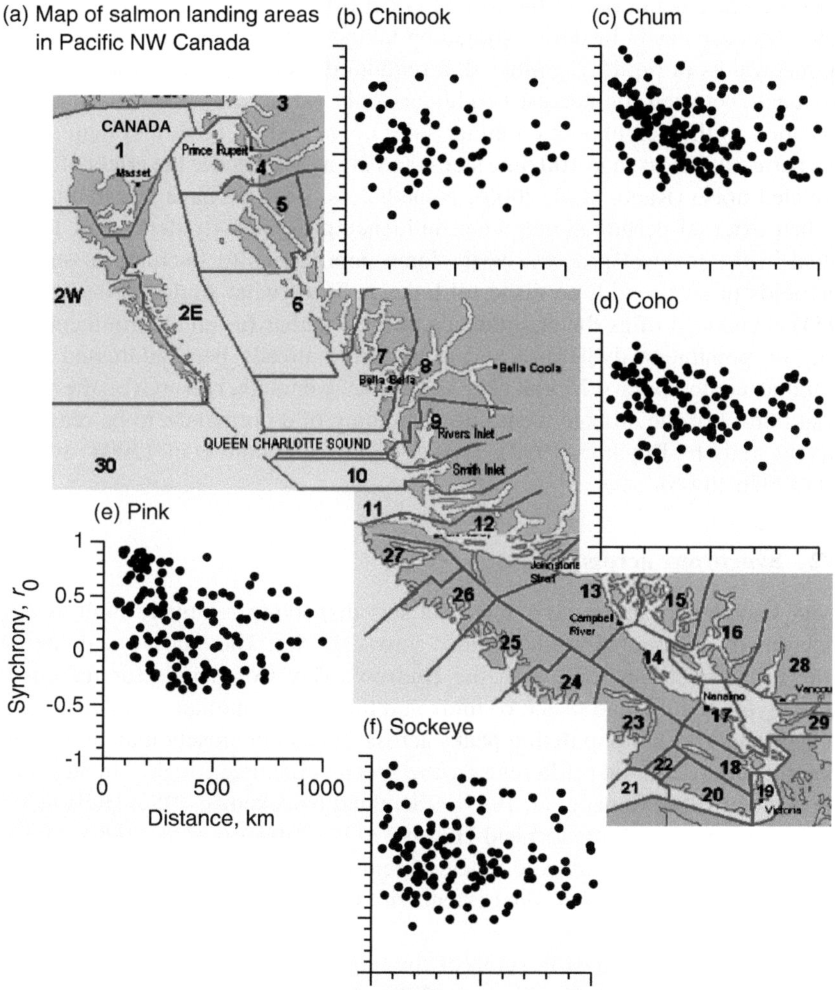

Figure 5.3. (a) Map of salmon landing areas in Pacific North-West Canada surrounding Queen Charlotte Island and Vancouver Island. The five insets are synchrony vs. distance scatter-plots for five salmon species. (b) Chinook salmon: data from landing areas (identified with numbers in the map), 1–8, 12, 20, 26, and 29, (c) Chum salmon: 1–8, 10–13, 20, 21, 25, 26, 27, and 29. (d) Coho salmon: 1–8, 10–12, 20, 21, 23, 25, 27, and 29. (e) Pink salmon: 1–8, 10–13, 20, 23–25 and 27. (f) Sockeye salmon: 1–8, 10–13, 18, 20, 23, 27, and 29. Note that the x- and y-axis scales are the same in all panels but for sparing space labelling is given only with panel (e).

Exercises like these highlight the problem that although synchrony among populations can be achieved through different mechanisms, synchrony alone – regardless of its decline with increasing distance – does not give much insight into its underlying causes. Nor can we always conclude that synchrony over a small scale is due to dispersal, while being due to the Moran effect over a larger scale (cf. Paradis et al. 1999). This can easily be demonstrated by taking 10,000 populations, with population renewal as in Eq. (5.2), initiated at random densities and that share dispersing individuals between the nearest neighbours only (in our example 5% of the focal population to the neighbours on either side, i.e. $m = 0.1$). Such a system synchronises perfectly in less than 100 time steps, and even faster under the effect of globally correlated noise (Ranta et al. 2006). Nonetheless, here we have re-established the fact that external perturbations, when influencing the density-dependent feedback system in the renewal process, easily force populations to fluctuate in step. This even holds in a case with no dispersal linkage. Thus, what Andrewartha and Birch (1954) envisioned of as the significance of the weather (as an environmental force acting on population dynamics) had at that time already been addressed by their Australian contemporary, Moran (1953). Curiously enough, research on the synchronising impact of the climate went into something of a dormancy, to be reawakened to lively study by Royama (1992), as reviewed by Liebhold et al. (2004) and Ranta et al.(1997b, 2006).

5.3.2 Synchrony across species

To our knowledge, there are a few data sets displaying relatively high degree of synchronicity in population fluctuations across species. The oldest one comes from Butler (1953) who showed, with the Hudson's Bay Company's fur records, that Canada lynx, red fox and fisher, or mink and muskrat as another set of species (Figure 5.4) tend to reach population peaks across Canada in closely matching years. It is interesting to note that different grouse species fluctuate roughly in step in Finland (Lindén 1988; Ranta et al. 1995b), Scotland (Mackenzie 1952; Hudson 1992), and the Italian Alps (Cattadori and Hudson 1999; Cattadori et al. 2000). The same is also true of Scandinavian vole species (Henttonen 1985; Korpimäki and Norrdahl 1998; Stenseth 1999), and to some extent with British aphids and moths (Hanski and Woiwod 1993).

It is not entirely clear why some of the taiga forest fauna display synchronous dynamics across species. Ranta et al. (2006) simply extended the Moran effect, a common forcing in the population renewal process, to cover a number of differing species. They showed that the Moran effect is a suitable candidate for synchronizing dynamics even across species differing in their density-dependent processes and life history. With some reservations, the general conclusion is that when the form of density dependence is mixed (increasing life history differences among the species), the synchronizing capacity of the Moran effect becomes less obvious. As long as the deterministic renewal processes in different species yield asymptotically stable dynamics (when populations are not disturbed), synchronicity across species via the Moran effect is possible even with different species' demographies (Ranta et al. 2006).

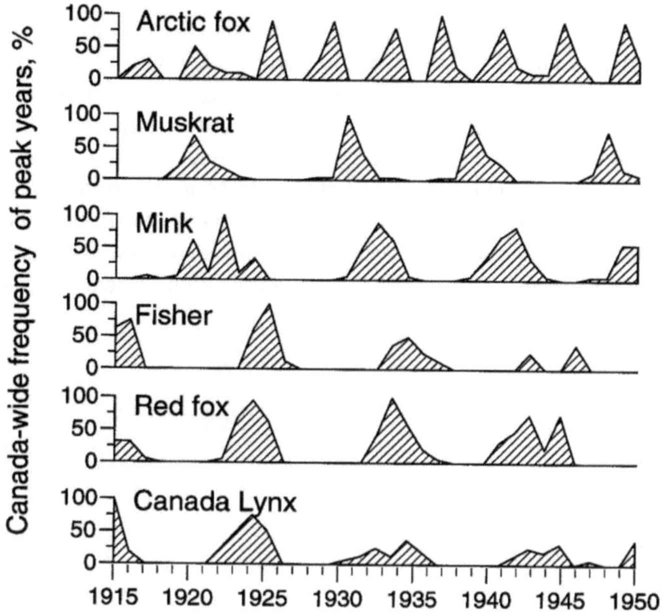

Figure 5.4. Nationwide frequency of peak years in population fluctuations of Canadian mammals (from data in Butler 1953). Butler divided Canada into 63 zones and scored from the Hudson's Bay Company's records in how many of the zones the different species had their population high year during the winters 1915–1916 to 1950–1951. For example, with perfect synchrony the Canada lynx peak years should all have coincided at perfectly matching 10-year intervals. That there is some spread (e.g. from 1921 to 1926 with the high in 1924) indicates that the synchrony is not perfect and that it may level off with increasing distance. Note also the pronounced match across species. For Butler (1953) these data suggested three clusters of species: Canada lynx, red fox and fisher; mink and muskrat and arctic fox.

An alternative explanation for the synchrony patterns found across species is that trophic interactions are of importance (Ydenberg 1987; Ims and Steen 1990; Korpimäki and Norrdahl 1991). Predator–prey (or host–parasitoid/host–disease) interactions act as surrogates for the Moran effect, by disturbing the density-dependent renewal process, hence yielding synchronicity (Ranta et al. 2006). However, this explanation calls for rather tight food web interactions to be valid across a large number of species, as all relevant interspecific interactions are incorporated across the community.

5.3.3 Population travelling waves

(Elton 1924; Elton and Nicholson 1942) first made the Hudson's Bay Company's fur records known to ecologists. Since then these data have been used repeatedly to demonstrate various aspects of predator–prey (Canada lynx – snowshoe hare) dynamics, and the anticipated causes of the pronounced 10-year periodicity in population fluctuations of these two species: Elton and Nicholson (1942) has been cited over 200 times in the ISI listed journals over the past 10 years alone. The Canada lynx dynamics (1919–1986, Royama 1992), when compared across the eight provinces in the dominion suggest a relatively high degree of overall synchrony (0.478 ± 0.104, mean ± 95% confidence limits; Figure 5.5a). To us, observing this high degree of synchrony over such a large geographical area (and using such robust methods) is still quite amazing (Ranta et al. 1997). More importantly, however, this long time series also renders it possible to examine the spatio-temporal dynamics in the temporal match of the lynx time series from various provinces. This was done by letting a sliding 15-year long time window pass through the data over consecutive

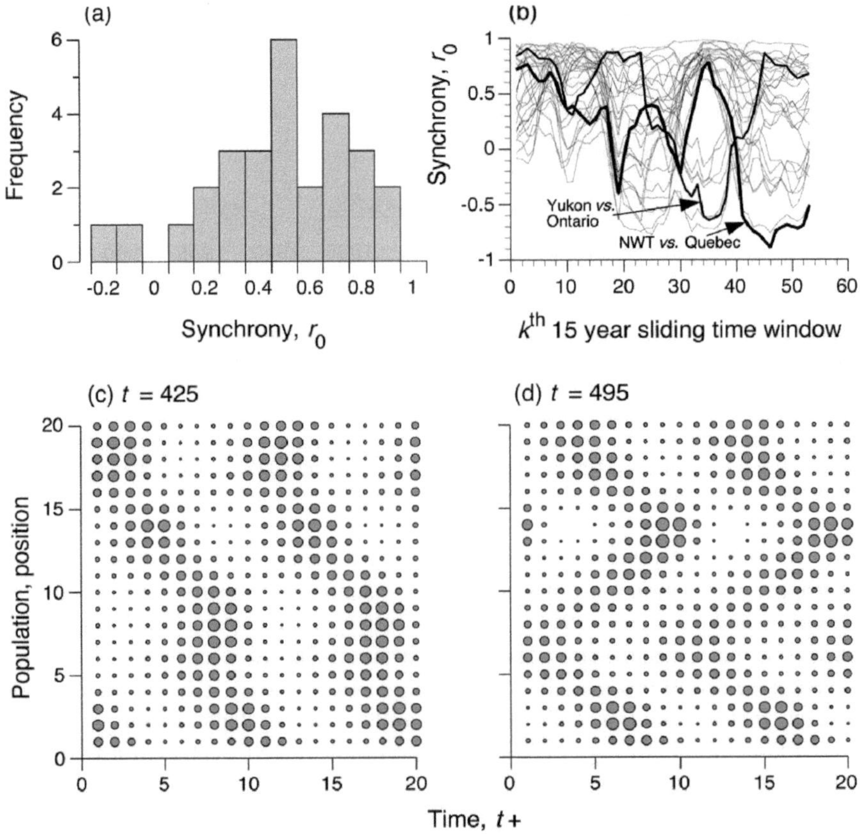

Figure 5.5. Canada lynx 1919–1986 population synchrony (a) when all time series (from 8 provinces) are compared in pairs, (b) and when a 15-year sliding time window is pushed step-by-step through the data (two province pairs are highlighted to indicate that synchrony comes and goes as time passes by). Two examples (c) and (d) of 20 time units (x-axis) at t = 425 and t = 495 of Canada lynx-like (10-year cyclic dynamics) in a selection of 20 central and linearly placed population subunits (y-axis; the sample populations arranged by their spatial proximity). The dot size refers to population size. The subpopulations are linked by 10% dispersal after negatively exponential kernel. (Modified from Ranta et al. 1997.)

years (Ranta et al. 1997c). Synchrony in pairs was calculated for each window. An apparent feature of the Canada lynx system is that, when compared in pairs of provinces, synchrony comes and goes as time passes by (Figure 5.5b).

This finding underlines the finding that over some time periods, a pair of lynx time series might be fluctuating in step, while at other times the temporal match can be less perfect. Such a system can be modelled by taking a second-order function for the lynx renewal process, a large number n of subpopulations in a coordinate space (so that in the middle of the space one can arrange 20 subunits in a row with a fixed distance from each other), letting global disturbance to influence all n units simultaneously and similarly and to allow negatively distance-dependent redistribution of individuals between breeding seasons (i.e. dispersal success is greater to nearby units than to faraway units). The system, when run over several repeated iterations (Ranta et al. 1997d) generates travelling waves in population density (Figure 5.5c,d). This describes the pattern commonly observed in numerous data sets: synchrony decaying against increasing distance. The spatio-temporal patterns arise as within patch peaks and troughs are propagated across the subpopulations in subsequent years, generating the impression that waves are passing through the landscape over time.

Several vole species annually cause severe damage in young sapling stands of pine, spruce, and birch in Finland. These are economically important forest trees; hence, the damage caused by voles has been recorded annually (Figure 5.6).

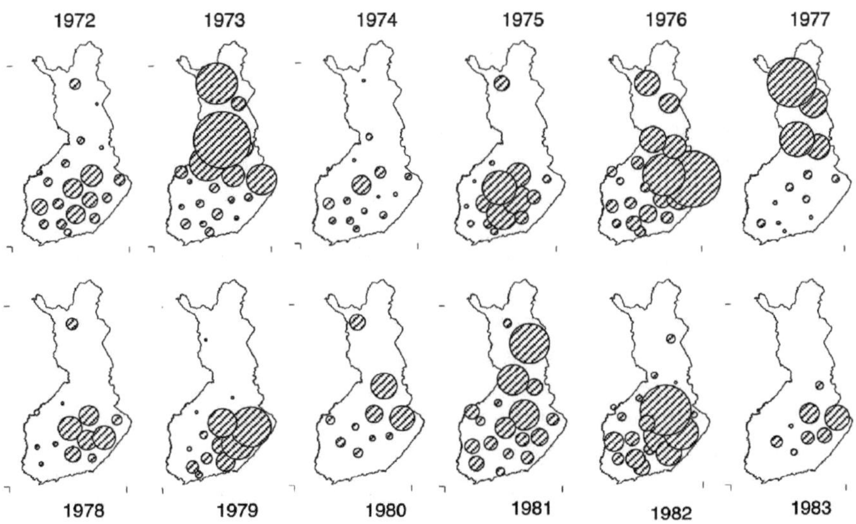

Figure 5.6. The vole damages in young forest tree plantations vary prominently over large areas between years in Finland. (Redrawn from Kaitala and Ranta 1998 using the original data.) The data, represent the number of seedlings destroyed giving an indication of annual local abundance of voles. The damage caused by voles to the tree stands is characterized by the 3–4-year periodicity, clustering of the damage, and the annual movement of the peak damage from one area to another creating a pattern of spatial asynchrony.

The records obtained from the different management districts in Finland suggest an interannual fluctuating pattern in the geographical distribution of peak damage areas (Ranta and Kaitala 1997; Kaitala and Ranta 1998). The damaged areas display 3- to 4-year cycles locally, and the region of peak damage changes between years. Thus, the temporal change of the location of the peak damage creates a dynamic pattern, where a certain degree of asynchrony from different areas can be observed. Locally, vole peaks are usually preceded by a 2- or 3-year period of population increase, which then ends with a sudden decline of the local population to extremely low levels (Hansson and Henttonen 1985).

A particularly nice example of periodic travelling waves in animal population dynamics is provided by the cyclic populations of Scottish field voles (Lambin et al. 1998; Bjørnstad et al. 1999; Mackinnon et al. 2001). Lambin and his associates have been studying vole population fluctuations since 1984 in the Kielder Forest and its surroundings. The data (1984–1998) show that field voles display a periodic travelling wave moving at the speed of ~20 km annually from West to East. Lambin and his associates are uncertain of the causes of the vole travelling wave, but they are adamant (Mackinnon et al. 2001) that nomadic predators and climatic factors are not responsible. This leaves us with demographic processes, redistribution of individuals and pathogens as possible explanations. Another Scottish example is the long-term data on red grouse population cycles in the Highlands (Moss et al. 2000). The authors discovered that the speed of the wave is 2–3 km annually and that it travels from the centre of the study area towards the margins. Moss et al. (2000) attribute demographic processes for the presence of this wave.

Larvae of the larch budmoth are major pests on larch forests in the European Alps. With periodic outbreaks at 8- to 10-year intervals they cause wide-ranging defoliation in larch forests, hence forest managers have kept a close eye on, and detailed records of, such outbreaks. This moth exhibits profound and characteristic periodic temporal oscillations moving like a wave through the Alps (Bjørnstad et al. 2002). Phases of population lows and highs in larch budmoth pass as waves through the forested areas of European Alps towards Northeastern Europe with a speed of about 200 km per year. When the spatial redistribution of individuals links nearby units together, aggregated areas in matching phase may emerge. Aggregates of population highs travel in the landscape as ripples in a pond. A travelling wave progressing through a region typically causes areas of population high and lows separated in space. When long-term population data from such areas are analysed, one finds that populations tend to fluctuate in synchrony and that the level of synchrony fades away with increasing distance, to increase again when another wave top is reached.

These are the classical travelling waves. A related, albeit conceptually different spatial–temporal phenomenon are the so-called travelling wavefronts – the spread of species into areas where they have not been found earlier. A well-documented example is the invasion of the muskrat in Europe after a few individuals escaped from a farm near Prague in 1905. Another classical example is the European starling: first individuals were released in New York's Central Park in 1880; by 1954 the species' range reached the Pacific. These are archetypes of a wavefront passing through the landscape in all directions wherever the habitat is suitable (Shikesada and Kawasaki 1997).

For obvious logistic reasons, population data with high-enough spatial resolution and long-enough temporal coverage to reveal travelling waves are rare. Nonetheless, ecologists have been able to demonstrate the presence of theory-anticipated travelling waves in real-world data. It is interesting that the best-documented examples of travelling waves in populations (Smith 1983; Ranta and Kaitala 1997; Kaitala and Ranta 1998; Lambin et al. 1998; Moss et al. 2000; Grenfell et al. 2001) originate from periodically fluctuating populations. Presently there is an ensemble of alternative views on the causes of such dynamics (Hudson et al. 1998; Stenseth 1999; Lindström et al. 2001; Fowler 2005). It remains to be seen whether periodic dynamics, regardless of their underlying mechanism, generate travelling waves more easily than dynamics of any other kind. Nonetheless, these results underline that the spatio-temporal process is a function of population renewal after a certain density-dependent structure (demography, or life history) spatial redistribution of individuals and the Moran effect (Ranta et al. 1997b).

5.4 INVISIBILITY OF ENVIRONMENTAL FACTORS

The so-called visibility problem refers to the challenge of understanding the underlying process from observed patterns in time-series data of population abundance (Ranta et al. 2000): *Is the fingerprint of environmental variability visible in the time series of population abundance?* Ranta et al. (2000) addressed the problem by allowing a deterministic population model to be affected by environmental noise. They searched for a correlation between the pure noise process and its manifestation in the population time series. Depending on the underlying form of population dynamics (whether the deterministic model was stable or not) and on the exact nature of the noise signal, the putative correlations are often masked.

Ranta et al. (2000) explored the visibility of external noise (structured noise with short- and long-term components) using Ricker dynamics as a model for population renewal. A high correlation was observed only in conditions where the equilibrium population size of the deterministic population dynamics was locally stable in the absence of external disturbance (Figure 5.7). As an extension, Lundberg et al. (2002) studied Ricker dynamics, split into births and deaths (Ripa and Lundberg 2000), as the skeleton of the population renewal process. The noise affected births, deaths, or both. Regardless of noise colour and which process it was influencing, the noise was again only visible when the underlying population dynamics were locally stable.

Kaitala and Ranta (2001) explored noise visibility using age-structured populations for iteroparous and semelparous breeders. The noise was never found to be visible with semelparous breeders. However, with iteroparous breeders, the noise was shown to be visible when the deterministic kernel of the renewal process

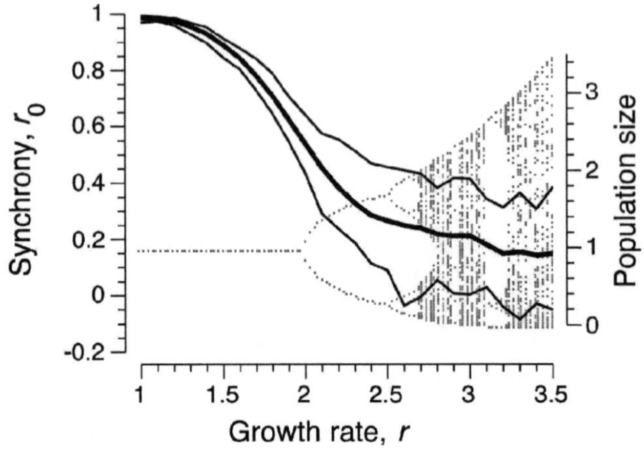

Figure 5.7. Synchrony (for the Ricker dynamics and the external noise (measured with cross correlation with lag -1). The solid line indicates average over 100 replicated runs, and the thinner lines include 95% of all observed correlation coefficients. Growth rate of the Ricker function r is given on the x-axis. (Modified from Ranta et al. 2000.) The shadow in the background is the bifurcation diagram for the Ricker dynamics.

produced locally stable population dynamics and when the noise influenced the youngest age group only. Otherwise, detection of the noise in the population signal was difficult.

Research by Laakso et al. (2001) and Greenman and Benton (2001) illustrated an additional problem. In terms of noise visibility, the situation becomes even worse if an assumption of non-linear responses to environmental variability is made. An environmental variable, e.g. temperature, rarely affects population renewal in a linear manner (e.g. Stenseth and Mysterud 2002). Rather, the effect is often dome-shaped or sigmoid (or even a binary all-or-none event). Should that be the case, the expected correlation between the environmental variable and the population dynamics we may use as a probe becomes even more compromised.

Hence, the time series of population change does not always necessarily carry any extractable information about the environmental variability that is affecting the population, unless complemented with auxiliary data about the environment and the focal population studied (Turchin 2003). The basic message is that the fingerprint of the external signal is very difficult to recognize within the noise-modulated dynamics of the focal population. It appears that noise is more easily visible in the noise-modulated population data if the underlying renewal process yields locally stable (not periodic or more complex) dynamics. This conclusion is not very specific to the details of the renewal kernel, highlighting the uncertainty about the demography–environment interaction. What are badly needed, therefore, are both solid *a priori* stochastic models of population renewal and, at least, well-informed guesses about the nature of the environmental variability affecting the demography of the population in question.

A related problem was explored by Jonzén et al. (2002), who modelled population dynamics as a linear process with two lags, termed an AR(2) process, modified by noise of known structure. We shall simplify their approach here, by using a first-order autoregressive process (Eq. (5.2) for a single population). This AR(1) process is the kernel of the population renewal process, with its single parameter, a_1, encapsulating the essence of life history (i.e. the biology of the target species) in the population renewal process. This dynamical process is also modulated by a noise signal. The task is to recover the underlying model structure given the data thus generated. The question is whether we can learn anything about the demographic (density-dependent) and environmental processes by estimating model parameters from a noisy time series?

The answer is that the noise signal is reasonably visible in the AR(1) process when the modulating noise has an autocorrelated structure, but not visible at all if the perturbation is white noise (Figure 5.8a,c,d). Attempting to recover biology from the noise-modulated population time-series data, by fitting an AR(1) statistical model to the simulated data for each of the parameter values used, indicates reasonably good recovery under white noise, when the noise is not detectable (Figure 5.8d). In fact, this simply proves that the AR(1) theory works well when the modulating signal is white noise, just as Box et al. (1994) and many other time-series analysts have noted. However, parameter recovery is far from perfect (except at the extremes) under blue or red noise (Figure 5.8b,f). Hence, when our visibility test says that the noise signal and the noise-modulated population signal have something in common (Figure 5.8a,e), we fail in recovering the underlying biology (as captured by the parameter a_1 of the AR(1) process). One can fine-tune the analysis by putting the noise term as covariate in general linearized model. We did this, and used Akaike information criterion to assess whether the noise-as-covariate model gave a better fit to the noise-modulated data than just the plain noise-modulated data. The answer was unambiguous, including noise as covariate does not enhance performance of the fitted model. Kaitala and Ranta (2006) give further details of this analysis.

In summary, the answer from the above analysis and from that by Jonzén et al. (2002) is that the skeleton of the population dynamics can only be recovered under very specific conditions. This underlines the “inverse” problem (Wood 1997), which means that it may be impossible to infer from pattern (the time series) the underlying process, i.e. the interaction between a revealed demography and a correctly identified environmental variability.

5.5 FUTURE DIRECTIONS

Half a century has elapsed since Patrick Moran outlined his ingenious ideas of how environmental perturbation is capable of synchronizing the dynamics of populations sharing a common environment (Moran 1953). His idea was in direct response to his

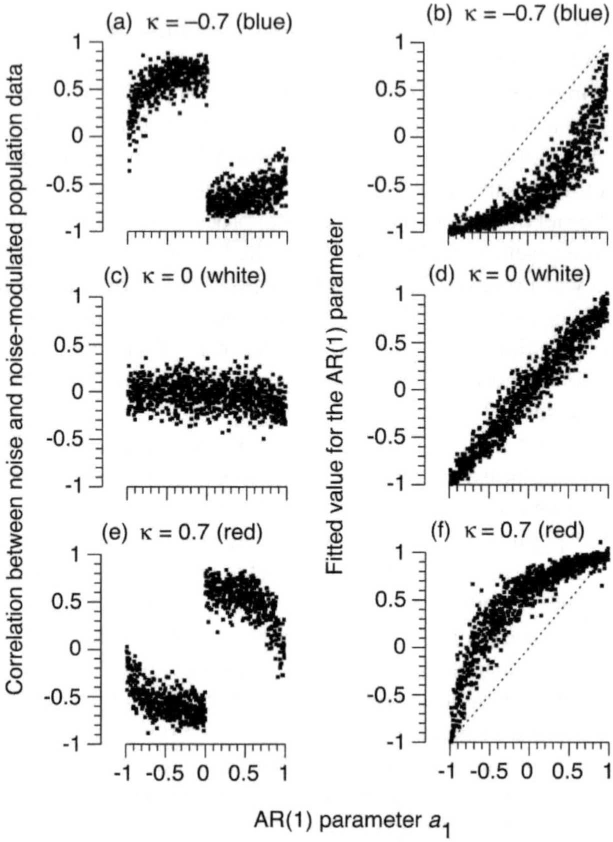

Figure 5.8. First-order linear autoregressive process, AR(1) is used as the kernel of population renewal. The process is disturbed with additive external noise (ranging between -1 and 1) having temporal structure [serial correlation being either $\kappa = -0.7$ (a,b), $\kappa = 0$ (c,d), or $\kappa = 0.7$ (e,f)]. Panels a, c and e give correlation (high positive and negative values are an indication of good visibility) between noise and noise-modulated population dynamics against various values of a_1, the AR(1) parameter. Panels (b), (d), and (f) give fitted values for the AR(1) parameter against the actual values of a_1 used (1:1 line indicated). Note that the discontinuity in panels (a) and (e) is simply due to the a_1 coefficient changing sign when passing zero.

interactions with Charles Elton, who initiated the research on large-scale phenomena in population dynamics (Elton 1924; Lindström et al. 2001). Now, we can attempt to answer the question: What have we learned about population synchronicity during the past five decades, or rather, during the three quarters of a century (Elton 1924). This question has already been subject of a few reviews (Ranta et al. 1997b, 2006; Hudson and Cattadori 1999; Bjørnstad et al. 1999b; Liebhold et al. 2004; Ranta et al. 2006), and one can conclude that by now we have much more data on the taxonomic and geographic extent of synchronicity both within and across species. We have an enriched palette with a number of agents, potentially forcing populations regulated by density-dependent feedback, to fluctuate in step: the Moran effect, redistribution of individuals between breeding seasons and trophic or disease interactions. These may act alone, but are more likely performing in concert with each other.

And yet, in most if not in all cases, more information is required before the underlying causes of population synchrony can be reliably identified for specific populations. This conclusion would of course feel like a real letdown if the current knowledge would not provide at least a stepping-stone forward. Therefore, we feel the pertinent question is: *What information is missing?* In the section that follows we shall make an attempt to construct a "shopping list". For any particular system, there is a need of higher quality data on the ranges of redistribution (through dispersal) of individuals. Basically, only anecdotal information is available for the dispersal distances taken by immature individuals from their natal to their breeding areas in most species used in the examples here, or elsewhere (Liebhold et al. 2004; Ranta et al. 2006). This shortfall is primarily due to a shortage of empirical data, however, most rewarding data collection can only be guided and motivated by solid theory.

Whether the theoretical advancement in spatial–temporal dynamics of populations has matured sufficiently will only be seen in the next few years. Presently, uncomfortable as it is, we have to say that research in the space-time dynamics of populations provides reasonably little useful guidance to give to empiricists. We can summarise this in layman terms somewhat provocatively as: "Go and collect population time series data in as many places as possible over as many matching years as possible. Collect also the best possible auxiliary data, in case somebody else a few decades later thinks those information would have been relevant and will make a good use of your efforts." However, this is clearly not a very satisfactory research programme, and better theory on the significance of dispersal (and various dispersal kernels) is called for. An encouraging example showing that progress is not entirely impossible is provided by Ims and Andreasson (2000) who used field experimentation to disentangle the significance of dispersal and predation of vole population synchrony.

We also lack the precise identity of the agents causing a Moran effect in most cases where it is called as an explanation. This is unfortunate, indeed! For just a few scattered cases we have a set of candidates that really might influence population synchrony. One of these is the Soay sheep (Grenfell et al. 1998; Clutton-Brock et al. 2004). Populations of this feral sheep fluctuate in temporal step on two geographically close, remote islands in the Atlantic off the West coast of Scotland. Due to their separation from each other by the sea, dispersal can be eliminated as the synchronizing agent. Therefore, a weather-induced Moran effect provides a general explanation. However, as the Soay sheep data set is long-term and individual-based, it provides a basis for very detailed analyses. Such analyses have shown that when attempting to understand the interactions between population fluctuations and climate, the devil really can be in the detail (Clutton-Brock and Coulson 2002). For instance, whether the Soay sheep population will crash or not depends not only on the winter weather conditions but also on the timing of adverse conditions, population size, and population structure, as density dependence affects different sexes and age-classes differently (Coulson et al. 2001; Hallett et al. 2004). These same studies also demonstrate that trying to identify one single factor equalling the Moran effect may be futile as it does not have to be the same factor in different years: increased winter mortality in the Soay sheep can be caused by high rainfall, high winds, or low temperatures.

With a data on red grouse populations in Northern England Cattadori et al. (2005) suggest that that climate affects trophic interactions and that this process could be a mechanism for synchronizing spatially distributed populations. In specific years the size of red grouse populations in northern England either increases or decreases in synchrony. In these years, widespread and correlated climatic conditions during May and July affect populations regionally and influence the density-dependent transmission of a gastrointestinal nematode, a parasite that reduces grouse fecundity. Dry and warm Mays followed by cold and wet Julys reduce parasite transmission (grouse go up in numbers), while wet Mays and warm Julys increased parasite transmission (grouse numbers go down). This study suggests that the effects of common climatic events influence the red grouse population renewal through a complex host and parasite interaction, the result being synchrony in grouse population fluctuations. We have to remain waiting for other studies on synchronicity with this high resolution of details to unravel the complexity of the true nature of Moran effect (so easy to implement into a model, as seen in Eq. (5.2)).

There are also some obstacles in gaining more thorough understanding of the travelling waves. First, despite the rapidly increasing number of the studies of travelling waves in populations, there seem not to exist a general and widely agreed understanding on what they are. Some studies hint to a single wavefront (high population peak), proceeding in space to a certain direction, documented in the context of an invasion of a species to new areas. Others argue about synchronously and repeatedly changing population sizes over a range of area, in which changes do not occur simultaneously in all places, but with some delay between neighbouring regions. Yet another discrepancy is the scale of such waves; some studies report waves within the scale of few kilometres, while others may look at the same phenomena at the scale of hundreds of kilometres. Also, the speed (not to mention about direction) of the wave may vary orders of magnitude. Second, it is an intriguing problem that the available statistical tools seem to be quite ineffective in tackling this issue.

Concerning the visibility problem, we face the discrepancy between theoretical advances and the limitations of inference using real data. With theoretical models, we can sometimes detect the effect of environmental factors on the population dynamics under rather strict assumptions, restricting the analysis to certain classes of population dynamics models and the qualities of environmental variability. The fundamental difficulty in analysing "real" population data is that we cannot really know whether our assumptions for either of these hold. Moreover, studies based on correlative analyses can be biased as the world is full of population records and other time series so that the possibility of spurious correlations is substantial.

To conclude, we feel that the next major steps in understanding spatial and temporal synchrony of populations hinge on improvements on both theory and data. Concerning theory, better analytical tools for disentangling between dispersal and external disturbances are needed. Challenging as this is, it is not impossible (e.g. Lande et al. 1999). However, there is still plenty of scope for development here before the theory and data can be matched smoothly. This could perhaps partly be aided by having a more realistic and detailed picture on dispersal processes in nature. Among the most important questions are: Who disperses and how far? What are the main determinants triggering dispersal decisions? In general, more detailed long-term data on population demography are needed; knowing which parts of the demography vary in time and space can also help in revealing the biological mechanisms behind population fluctuations (e.g. Oli and Armitage 2004). With such data one is better armed to address questions that Elton (1924), (Andrewartha and Birch 1954; see also Yodzis 1989) raised, and to make progress along the way Moran (1953) envisioned.

5.6 LITERATURE CITED

Andrewartha, H.G. and L.C. Birch. 1954. *The distribution and abundance of animals.* The University of Chicago Press, Chicago.

Barbraud, C. and H. Weimerskirch. 2003. Climate and density shape population dynamics of a marine top predator. *Proceedings of the Royal Society of London, Series B* 270: 2111–2116.

Bjørnstad, O.N., R.A. Ims, and X. Lambin. 1999a. Spatial population dynamics: analysing patterns and processes of population synchrony. *Trends in Ecology and Evolution* 14: 427–432.

Bjørnstad, O.N. and B. Bolker. 2000. Canonical functions for dispersal-induced synchrony. *Proceedings of the Royal Society of London, Series B.* 267: 1787–1794.

Bjørnstad, O.N., M. Peltonen, A.M. Liebhold, et al. 2002. Waves of larch budmoth outbreaks in the European Alps. *Science* 298: 1020–1023.

Bodenheimer, F.S. 1938. *Problems of animal ecology*. Oxford University Press, Oxford.

Box, G.E.P., G.M. Jenkins, and G.C. Reinsel. 1994. *Time series analysis: forecasting and control.* Prentice-Hall, Englewood Cliffs, NJ.

Butler, L. 1953. The nature of cycles in populations of Canadian mammals. *Canadian Journal of Zoology* 31: 242–262.

Cappuccino, N. and P.W. Price. 1995. *Population dynamics: new approaches and synthesis.* Academic Press, London.

Cattadori, I.M. and P.J. Hudson. 1999. Temporal dynamics of grouse populations at the southern edge of their distribution: the Dolomitic Alps. *Ecography* 22: 374–387.

Cattadori, I.M., S. Merler, and P.J. Hudson. 2000. Searching for mechanisms of synchrony in spatially structured gamebird populations. *Journal of Animal Ecology* 69: 620–638.

Cattadori, I.M., D.T. Haydon, and P.J. Hudson. 2005. Parasites and climate synchronize red grouse populations. *Nature* 433: 737–741.

Clutton-Brock, T.H., B.T. Grenfell, T. Coulson, A.D.C. MacColl, A.W. Illius, M.C. Forchhammer, K. Wilson, J. Lindström, M.J. Crawley, and S.D. Albon. 2004. Population dynamics in Soay sheep. In: *Soay sheep: dynamics and selection in an island population* (eds., T.H. Clutton-Brock and J.M. Pemberton), pp. 52–88. Cambridge University Press, Cambridge.

Clutton-Brock, T.H. and T. Coulson. 2002. Comparative ungulate dynamics: the devil is in the detail. *Philosophical Transactions of the Royal Society of London, Series B* 357: 1285–1298.

Coulson, T., E.A. Catchpole, S.D. Albon, B.J.T. Morgan, J.M. Pemberton, T.H. Clutton-Brock, M.J. Crawley, and B.T. Grenfell. 2001. Age, sex, density, winter weather, and population crashes in Soay sheep. *Science* 292: 1528–1531.

Elton, C.S. 1924. Periodic fluctuations in the numbers of animals: their causes and effects. *British Journal of Experimental Biology* 2: 119–163.

Elton, C.S. and M. Nicholson. 1942. The ten-year cycle in numbers of the lynx in Canada. *Journal of Animal Ecology* 11: 215–244.

Engen, S. and B.E. Sæther. 2005. Generalizations of the Moran effect explaining spatial synchrony in population fluctuations. *American Naturalist* 166: 603–612.

Fowler, M. 2005. Interactions between maternal effects and dispersal. *Oikos* 110: 81–90.

Greenman, J.V. and T.G. Benton. 2001. The impact of stochasticity on the behaviour of nonlinear population models: synchrony and the Moran effect. *Oikos* 93: 343–351.

Grenfell, B.T., O.N. Bjørnstad, and J. Kappey. 2005. Travelling waves and spatial hierarchies in measles epidemics. *Nature* 414: 716–723.

Grenfell, B.T., K. Wilson, B.F. Finkenstädt, T.N. Coulson, S. Murray, S.D. Albon, J.M. Pemberton, T.H. Clutton-Brock, and M.J. Crawley. 1998. Noise and determinism in synchronised sheep dynamics. *Nature* 394: 674–677.

Haeckel, E. 1870. Über Entwickelungsgang und Aufgabe der Zoologie. *Jenaische Zeitschrift* 5: 353–370.

Hanski, I. and I.P. Woiwod. 1993. Spatial synchrony in the dynamics of moth and aphid populations. *Journal of Animal Ecology* 62: 656–668.

Hallett, T.B., T. Coulson, J.G. Pilkington, T.H. Clutton-Brock, J.M. Pemberton, and B.T. Grenfell. 2004. Why large-scale climate indices seem to predict ecological processes better than local weather. *Nature* 430: 71–75.

Hansson, L. and H. Henttonen. 1985. Gradients in density variations of small rodents: the importance of latitude and snow cover. *Oecologia* 67: 394–402.

Henttonen, H. 1985. Predation causing extended low densities in microtine cycles: further evidence from shrew dynamics. *Oikos* 45: 156–157.

Holmgren, M., M. Scheffer, E. Ezcurra, J.R. Gutierrez, and G.M.J. Mohren. 2001. El Niño effects on the dynamics of terrestrial ecosystems. *Trends in Ecology and Evolution* 16: 89–94.

Hudson, P.J. 1992. *Grouse in space and time.* Came Conservancy Trust, Fordingbridge.

Hudson, P.J., A.P. Dobson, and D. Newborn. 1998. Prevention of population cycles by parasite removal. *Science* 282: 2256–2258.

Ims, R.A. and H. Steen. 1990. Geographical synchrony in microtine population cycles: a theoretical evaluation of the role of the nomadic avian predators. *Oikos* 57: 381–387.

Ims, R.A. and H.P. Andreassen. 2000. Spatial synchronization of vole population dynamics by predatory birds. *Nature* 408: 194–196.

Jaksic, F.M. 2001. Spatiotemporal variation patterns of plants and animals in San Carlos de Apoquindo, Central Chile. *Revista Chilena de Historia Natrual* 74: 477–502.

Jonzén, N., P. Lundberg, E. Ranta, and V. Kaitala. 2002a. The irreducible uncertainty of the demography–environment interaction in ecology. *Proceedings of the Royal Society of London, Series B* 269: 221–226.

Jonzen, N., A. Hedenstroem, C. Hjort, Å. Lindström, P. Lundberg, and A. Andersson. 2002b. Climate patterns and the stochastic dynamics of migratory birds. *Oikos* 97: 329–336.

Kaitala, V. and E. Ranta. 1998. Travelling wave dynamics and self-organization in a spatio-temporally structured population. *Ecology Letters* 1: 186–192.

Kaitala, V. and E. Ranta. 2001. Is the impact of environmental noise visible in the dynamics of age-structured populations? *Proceedings of the Royal Society of London, Series B* 268: 1769–1774.

Kaitala, V. and E. Ranta. 2006. Environmental noise and population renewal process. *Journal of Negative Results*, in press.

Korpimäki, E. and K. Norrdahl. 1991. Do breeding nomadic avian predators dampen population fluctuations of small mammals? *Oikos* 62: 195–208.

Korpimäki, E. and K. Norrdahl. 1998. Experimental reduction of predators reverses the crash phase of small-rodent cycles. *Ecology* 79: 2448–2455.

Laakso, J., V. Kaitala, and E. Ranta. 2001. How does environmental variation translate into biological processes? *Oikos* 92: 119–122.

Lack, D. 1954. *The natural regulation of animal numbers*. Oxford University Press, New York.

Lambin, X., D.A. Elston, S.J. Petty, and J.L. MacKinnon. 1998. Spatial asynchrony and periodic travelling waves in cyclic populations of field voles. *Proceedings of the Royal Society of London, Series B* 265: 1491–1496.

Lande, R., S. Engen, and B.E. Sæther. 1999 Spatial scale of population synchrony: environmental correlation versus dispersal and density regulation. *American Naturalist* 154: 271–281.

Liebhold, A., W.D. Koenig, and O.N. Bjørnstad. 2004. Spatial synchrony in population dynamics. *Annual Review of Ecology, Evolution, and Systematics* 35: 467–490.

Lindén, H. 1988. Latitudinal gradients in predator–prey interactions, cyclicity and synchronism in voles and small game populations in Finland. *Oikos* 52: 341–349.

Lindström, J., E. Ranta, H. Kokko, P. Lundberg, and V. Kaitala. 2001. From arctic lemmings to adaptive dynamics: Charles Elton's legacy in population ecology. *Biological Reviews* 76: 129–158.

Lundberg, P., J. Ripa, V. Kaitala, and E. Ranta. 2001. Visibility of demography-modulating noise in population dynamics. *Oikos* 96: 379–382.

Mackenzie, J.M.D. 1952. Fluctuations in the numbers of British tetraonids. *Journal of Animal Ecology* 21: 128–153.

Mackinnon, J.L., S.J. Petty, D.A. Elston, C.J. Thomas, T.N. Sherratt, and X. Lambin. 2001. Scale invariant spatio-temporal patterns of field vole density. *Journal of Animal Ecology* 70: 101–111.

Moran, P.A.P. 1953. The statistical analysis of the Canadian lynx cycle. II. Synchronization and meteorology. *Australian Journal of Zoology* 1: 291–298.

Moss, R., D.A. Elston, and A. Watson. 2000. Spatial asynchrony and demographic travelling waves during Red Grouse population cycles. *Ecology* 81: 981–989.

Murray, B.G. 1982: On the meaning of density dependence. *Oecologia* 53: 370–373.

Oli, M.K. and K.B. Armitage. 2004. Yellow-bellied marmot population dynamics: demographic mechanisms of growth and decline. *Ecology* 85: 2446–2455.

Ottersen, G., B. Planque, A. Belgrano, E. Post, P.C. Reid, and N.C. Stenseth. 2001. Ecological effects of the North Atlantic Oscillation. *Oecologia* 128: 1–14.

Paradis, E., S.R. Baillie, W.J. Sutherland, and R.D. Gregory. 1999. Dispersal and spatial scale affect synchrony in spatial population dynamics. *Ecology Letters* 2: 114–120.

Post, E., M.C. Forchhammer, N.C. Stenseth, and T.V. Callaghan. 2001. The timing of life-history events in changing climate. *Proceedings of the Royal Society London B* 268: 15–23.

Ranta, E and V. Kaitala, V. 1997. Travelling waves in vole population dynamics. *Nature* 390: 456.

Ranta, E., P. Lundberg, and V. Kaitala. 2006. *Ecology of populations*. Cambridge University Press, Cambridge.

Ranta, E., V. Kaitala, J. Lindström, and H. Lindén. 1995a. Synchrony in population dynamics. *Proceedings of the Royal Society London, Series B* 262: 113–118.

Ranta, E., J. Lindström, and H. Lindén. 1995b. Synchrony in tetraonid population dynamics. *Journal of Animal Ecology* 64: 767–776.

Ranta, E., J. Lindström, V. Kaitala, H. Kokko, H. Lindén, and E. Helle. 1997a. Solar activity and hare dynamics – a cross-continental comparison. *American Naturalist* 149: 765–775.

Ranta, E., V. Kaitala, and J. Lindström. 1997b. Spatial dynamics of populations. In: *Modelling spatiotemporal dynamics in ecology* (eds., J. Bascompte and R.V. Solé), pp. 47–62. Springer, Austin, TX.

Ranta, E., V. Kaitala, and P. Lundberg. 1997c. Population variability in space and time: the dynamics of synchronous population fluctuations. *Science* 278: 1621–1623.

Ranta, E., V. Kaitala, and J. Lindström. 1997d. Dynamics of Canadian lynx populations in space and time. *Ecography* 20: 454–460.

Ranta, E., V. Kaitala, and J. Lindström. 1999. Spatially autocorrelated disturbances and patterns in population synchrony. *Proceedings of the Royal Society of London, Series B* 266: 1851–1856.

Ranta, E., P. Lundberg, V. Kaitala, and J. Laakso. 2000a. Visibility of the environmental noise modulating population dynamics. *Proceedings of the Royal Society of London, Series B* 267: 1851–1856.

Ripa, J. and P. Lundberg. 2000. The route to extinction in variable environments. *Oikos* 90: 89–96.

Royama, T. 1992. *Analytical population dynamics*. Chapman & Hall, New York.

Royama, T. 2005. Moran effect on nonlinear population processes. *Ecological Monographs* 75: 277–293.

Sæther, B.E., J. Tufto, S. Engen, K. Jerstad, O.W. Rostad, and J.E. Skatan. 2000. Population dynamical consequences of climate change for a small temperate songbird. *Science* 287: 854–856.

Sinclair A.R.E., J.M. Gosline, G. Holdsworth, C.J. Krebs, S. Boutin, J.N.M. Smith, R. Boonstra, and M. Dale. 1993. Can the solar-cycle and climate synchronize the snowshoe hare cycle in Canada – evidence from tree rings and ice cores. *American Naturalist* 141: 173–198.

Shikesada, N. and K. Kawasaki. 1997. *Biological invasions: theory and practice.* Oxford University Press, Oxford.

Smith, C.H. (1983). Spatial trends in Canadian snowshoe hare, *Lepus americanus*, population cycles. *Canadian Field-Naturalist* 97: 151–160.

Stenseth, N.C. 1999. Population cycles in voles and lemmings: density dependence and phase dependence in a stochastic world. *Oikos* 87: 427–461.

Stenseth, N.C., K.-S. Chan, H. Tong, R. Boonstra, S. Boutin, C.J. Krebs, E. Post, M. O'Donoghue, N.G. Yoccoz, M.C. Forchhammer, and J.W. Hurrell. 1999. Common dynamic structure of Canada lynx populations within three climatic regions. *Science* 285: 1071–1073.

Stenseth, N.C. and A. Mysterud. 2002: Climate, changing phenology, and other life history traits: Nonlinearity and match-mismatch to the environment. *Proceedings of the National Academy of Sciences of the USA* 99: 13379–13381.

Turchin, P. 2003. *Complex population dynamics: a theoretical/empirical synthesis.* Princeton University Press, Princeton, NJ.

Wood, S.N. 1997. Inverse problems and structured-population dynamics. In: *Structured Population Models in Marine, Terrestrial, and Freshwater Systems* (eds., S. Tuljapurkar and H. Caswell). Chapman & Hall, London.

Ydenberg, R.C. 1987. Nomadic predators and geographical synchrony in microtine population cycles. *Oikos* 50: 270–272.

Yodzis, P. 1989. *Introduction to Theoretical Ecology.* Harper & Row, New York.

CHAPTER 6

INTERACTION ASSESSMENTS IN CORRELATED AND AUTOCORRELATED ENVIRONMENTS

JÖRGEN RIPA*

Department of Theoretical Ecology, Ecology Building, Lund University, SE-223 62 Lund, Sweden
**E-mail: jorgen.ripa@teorekol.lu.se, Phone: +46 (0)46 2223770, Fax: +46 (0)46 2224716*

ANTHONY R. IVES

Department of Zoology, UW-Madison, Madison, WI 53706, USA

6.1 ABSTRACT

Natural food webs are embedded in a variable environment, which causes population densities to fluctuate, despite a potential stable equilibrium. Population interactions as well as the characteristics of the environmental fluctuations determine

David A. Vasseur and Kevin S. McCann, The Impact of Environmental Variability on Ecological Systems, 111–131. © 2007 *Springer.*

the resulting population dynamics. Populations sensitive to the same kind of environmental disturbances will show correlated responses in their respective growth rates. Such "environmental correlation" between species can have profound effects on the populations' dynamics, e.g. generating a positive correlation between the abundances of two competitors, which makes a direct correlation a highly inappropriate measure of population interactions. However, multivariate time-series analysis will still identify and quantify population interactions correctly. The picture is more complicated if the environmental fluctuations are correlated over time – environmental autocorrelation causes biases in interaction assessments and possibly falsely identified delayed interactions. We present approximate expressions for the estimation bias, which show that the bias is the weakest when food web dynamics are close to unstable. In the absence of close to unstable dynamics the only way to avoid this estimation error is to incorporate the most important environmental drivers as covariates in the time-series analysis.

Keywords: food web dynamics, community dynamics, environmental stochasticity, correlation, autocorrelation, interaction assessment, multivariate time-series analysis.

6.2 INTRODUCTION

Field measurements of species interactions often requires large-scale manipulations of wild populations, such as PULSE or PRESS experiments (Bender et al. 1984). An alternative approach is to analyze time-series data of the interacting populations, utilizing the natural variability of population sizes (e.g. Seifert and Seifert 1976; Berryman 1991). The time-series analysis approach requires long time series, and therefore long-term projects, but is often less laborious than large-scale field experiments and ideally does not perturb the system from its natural conditions. The underlying assumption is that interactions between populations have a measurable impact on natural population dynamics. At the same time, the importance of environmental fluctuations for population dynamics has been long appreciated among ecologists (e.g. Elton 1924). A variable, stochastic, environment will continuously disturb an ecological system from any deterministic path, creating a stochastic process – a mixture of deterministic ecological processes and stochastic fluctuations. The dynamics of any population are thus determined by the nature of its intra- and interspecific interactions as well as the properties of the environmental fluctuations (Roughgarden 1975a,b; Royama 1981). Given that environmental fluctuations are important for population dynamics it is fair to ask to what extent they influence our ability to measure population interactions from time-series data. We will here study the possible effects of two important properties of a variable environment: its temporal autocorrelation and the correlation between species.

Steele (1985) and Pimm and Redfearn (1988) pointed out that many abiotic and biotic environmental factors may change slowly over time, thereby creating temporal autocorrelation in the stochastic fluctuations of populations' growth rates. Later theoretical work has shown the importance of environmental autocorrelation for the dynamics of single populations (Roughgarden 1975a; Strebel 1985; Ripa and Lundberg 1996; Kaitala et al. 1997; Petchey et al. 1997; Halley and Kunin 1998; Heino 1998; Cuddington and Yodzis 1999; Griebeler and Gottschalk 2000), as well as for interacting populations (Caswell and Cohen 1995; Ives 1995; Ripa et al. 1998; Ripa and Ives 2003). Further, an autocorrelated environment changes the dynamics of a population such that false time-delayed density-dependent interactions might be detected (Roughgarden 1975a; Royama 1981; Walters 1990; Williams and Liebhold 1995; Jonzén et al. 2002). Understanding the effects of environmental autocorrelation on population dynamics is consequently important for interpretation of time-series data, but its implications for the assessments of population interactions is to our knowledge not analysed before.

Interacting populations presumably live in the same habitat, which makes them face the same environment. If they respond strongly to the same kind of environmental fluctuations (such as temperature or precipitation), their growth rates will fluctuate in a correlated fashion. For convenience, we will describe two species as having "correlated environments" when we actually mean they show correlated responses to environmental fluctuations. The resulting correlation can be positive, if they respond in similar ways to similar fluctuations. It can also be negative, if they have non-overlapping environmental niches, which makes a good environment for one species a bad environment for the other (cf. Lehman and Tilman 2000). It seems plausible that such environmentally induced correlations in population growth rates may be mistaken for population interactions. It has been shown that the impacts on population dynamics from environmental correlation can be substantial (Roughgarden 1975b; Ripa and Ives 2003), but we know of no study on its implications for interaction assessments.

Here, we study the effects of first environmental correlation and second environmental autocorrelation on population interaction assessment. Each part starts with a short review of relevant earlier results on the dynamics of food webs subject to environmental variability.

We argue that time-series analysis is necessary to determine population interactions from density data alone, and that environmental data should be included to avoid falsely identified interactions and biases in assessing strengths of interactions. We also provide mathematical expressions for the bias created by an autocorrelated environment, should environmental data not be included in the time-series analysis.

6.3 FOOD WEB DYNAMICS IN CORRELATED ENVIRONMENTS

In a previous paper we (Ripa and Ives 2003) analyzed food web dynamics using a simple but general model of a food web in a stochastic environment:

$$\mathbf{x}(t) = \mathbf{B}\mathbf{x}(t-1) + \varepsilon(t), \tag{6.1}$$

where $\mathbf{x}(t)$ is a column vector with n elements representing population densities of n species at time t, $\mathbf{B}$ is a constant $n \times n$ matrix encapsulating interactions between populations, and $\boldsymbol{\varepsilon}(t)$ is a vector of environmental disturbances, one for each population. Population densities $\mathbf{x}(t)$ are standardized by subtracting the mean density of each species, such that the mean of $\mathbf{x}$ is a vector of zeros. The matrix $\mathbf{B}$ represents the community matrix for a discrete-time model. Thus, the element b_{ij} gives the per capita effect of species j on the change in the population of species i from time $t-1$ to time t. The elements of $\mathbf{B}$ can be arbitrarily chosen to encapsulate any ecological interaction, albeit excluding interactions with longer time lags than one unit time-step.

Environmental fluctuations given by vectors $\boldsymbol{\varepsilon}(t)$ have zero mean and are by assumption serially independent; thus, values of one element of $\boldsymbol{\varepsilon}(t)$ at two different times t and s, $\varepsilon_i(t)$ and $\varepsilon_i(s)$, are independent. Although there is no serial correlation, correlations between the environments of different species are allowed, such that $\varepsilon_i(t)$ and $\varepsilon_j(t)$ may have non-zero correlation, positive or negative. The environmental correlations as well as variances are summarized in the environmental covariance matrix Σ_ε, where a diagonal element $\sigma_{\varepsilon,ii}$ gives the environmental variance of population i and an off-diagonal elements $\sigma_{\varepsilon,ij}$ gives the covariance between the environments of species i and j.

The stochastic system defined in Eq. (6.1) is potentially complex with many interacting populations. However, much simplicity is gained by making a change in coordinates to "eigenvector coordinates". Each $n \times n$ matrix has (with few exceptions) n eigenvectors and n eigenvalues, one per eigenvector. The eigenvectors are (again with few exceptions) linearly independent and therefore make up an alternative coordinate system, where the eigenvectors constitute coordinate axes. By transforming the density vector $\mathbf{x}(t)$ to eigenvector coordinates $\mathbf{y}(t)$ the dynamic system (6.1) is transformed to:

$$\mathbf{y}(t) = \mathbf{J}\mathbf{y}(t-1) + \boldsymbol{\omega}(t), \tag{6.2}$$

where $\boldsymbol{\omega}(t)$ are the similarly transformed environmental disturbances $\boldsymbol{\varepsilon}(t)$. The matrix $\mathbf{J}$ has the eigenvalues of $\mathbf{B}$ on its diagonal and is otherwise zero, which means each component of the transformed vector $\mathbf{y}(t)$ has its own dynamics according to:

$$y_i(t) = \lambda_i y_i(t-1) + \omega_i(t) \tag{6.3}$$

where λ_i is the eigenvalue corresponding to eigenvector i. Consequently, the eigenvalues determine the dynamics along each eigenvector. A negative eigenvalue gives alternating, back and forth, dynamics in the direction in phase space given by the corresponding eigenvector. A positive eigenvalue, on the other hand, gives slowly changing, positively autocorrelated dynamics along the given direction. The variance along a given eigenvector i is given by (Ripa and Ives 2003):

$$\nu_{y,ii} = \frac{\sigma_{\omega,ii}}{1 - |\lambda_i|^2} \tag{6.4}$$

where $\sigma_{\omega,ii}$ is the transformed environmental variance along the same eigenvector. It follows that eigenvalues with an amplitude close to one give strong, high-amplitude fluctuations, at least compared to the environmental variance in the same direction.

Figure 6.1 illustrates the phenomena discussed above with a Lotka–Volterra model with two competing populations. The model has two components (populations) and thus the system matrix (after linearization) is a 2×2 matrix with two eigenvectors and two eigenvalues. We have chosen parameter values such that there is one eigenvector ($\mathbf{v}_1$) with a strong negative eigenvalue ($\lambda_1 = -0.8$) and another eigenvector with an equally strong (equal magnitude) positive eigenvalue ($\lambda_2 = 0.8$). The system dynamics are a combination of rapid fluctuations in one direction, corresponding

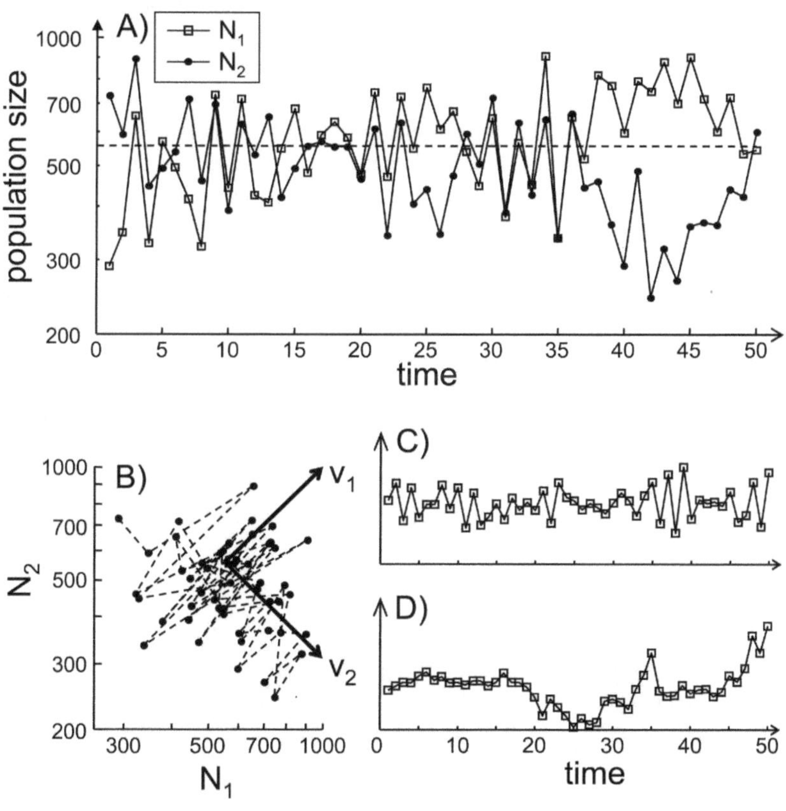

Figure 6.1. Simulated time series of a discrete-time competition model: $N_{1,t+1} = N_{1,t}\exp(r(1 - (N_{1,t} + \alpha N_{2,t})/K) + \varepsilon_{1,t})$, $N_{2,t+1} = N_{2,t}\exp(r(1 - (N_{2,t} + \alpha N_{1,t})/K) + \varepsilon_{2,t})$, where the $\varepsilon_{i,t}$ are serially uncorrelated normal random deviates with mean zero and variance 0.2^2. The system was first simulated for 100 time steps and the following 50 time steps are depicted here. Other parameter values: $r = 1.8$, $\alpha = 0.8$, $K = 1{,}000$, $\rho_{\varepsilon,12} = \text{corr}(\varepsilon_1(t), \varepsilon_2(t)) = 0$. (A) Population sizes of the two competitors over time. (B) The same simulation in phase space. (C, D) Projection of the simulated data onto the first (C) and second (D) eigenvectors as indicated in (B). The scale on the y-axis is arbitrary.

to the negative eigenvalue, and slow dynamics in the other direction, corresponding to the positive eigenvalue. Figure 6.1A depicts simulated time-series, also shown in phase space in Figure 6.1B. The dynamics in phase space can be projected onto the two eigenvectors (indicated in Figure 6.1B), resulting in oscillatory dynamics along the first eigenvector (Figure 6.1C) and highly autocorrelated dynamics along the second eigenvector (Figure 6.1D).

For further technical details, as well as treatment of complex eigenvalues, i.e. population cycles, see Reinsel (1997) and Ripa and Ives (2003).

6.3.1 Dynamical effects of correlated environments

According to Eq. (6.2), the variance along a given eigenvector depends not only on the corresponding eigenvalue (λ_i) but also on the variance of the transformed environmental fluctuations along the same eigenvector ($\sigma_{\omega,ii}$). The environmental variance along a given eigenvector depends on both the species-specific environmental variances (σ_{ii}), but also the pattern of correlations between them (ρ_{ij}). In the example of two competitors (Figure 6.1) a positively correlated environment means strong environmental variance along the first eigenvector ($\mathbf{v}_1$ in Figure 6.1B) but weaker variance along the second eigenvector ($\mathbf{v}_2$). The population dynamics change accordingly (Figure 6.2), such that population densities become positively correlated (Figure 6.2A) but also the dynamics become more dominated by rapid boom-and-bust dynamics corresponding to the second, negative, eigenvalue (Figure 6.2B). In conclusion, environmental correlation can change correlations between population densities in a quite straightforward way, but also has consequences for the individual population dynamics, i.e. the temporal autocorrelation of any individual population. As expected, a negatively correlated environment of the two competitors gives negatively correlated population densities (Figure 6.3A), but also population dynamics dominated by slow, positively autocorrelated dynamics (Figure 6.3B).

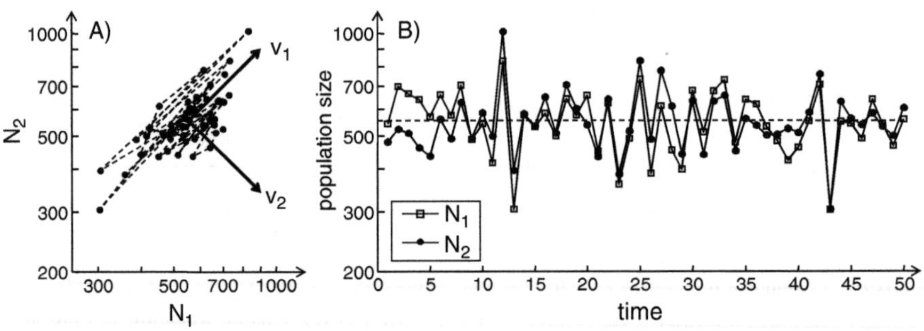

Figure 6.2. Phase portrait (A) and time series (B) of a computer simulation of the same competition model as was used in Figure 6.1, except $\rho_{\varepsilon,12} = \text{corr}(\varepsilon_1(t), \varepsilon_2(t)) = 0.9$.

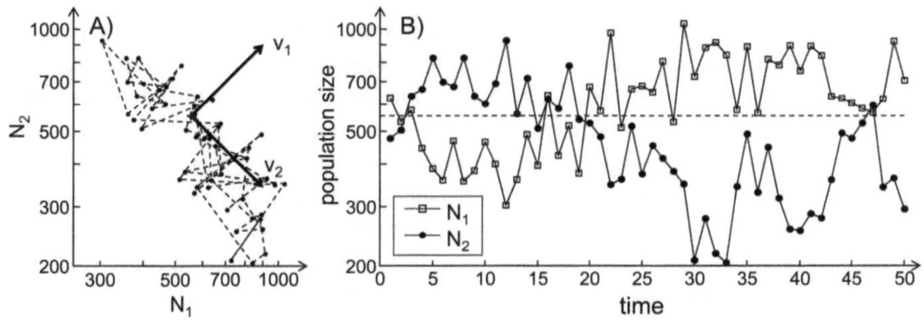

Figure 6.3. Phase portrait (A) and time series (B) of a computer simulation of the same competition model as was used in Figure 6.1, except $\rho_{\varepsilon,12} = \text{corr}(\varepsilon_1(t), \varepsilon_2(t)) = -0.9$.

To summarize this section, stochastic food web dynamics can be understood in terms of dynamics along the eigenvectors of the system matrix. As a rule of thumb, food web dynamics will be dominated by fluctuations along the eigenvectors with large magnitude eigenvalues. However, the variance and correlation pattern of the environmental fluctuations can also have an impact on which kind of population fluctuations will dominate the food web.

6.4 INTERACTION ASSESSMENTS IN CORRELATED ENVIRONMENTS

Environmental correlation presents challenges for any attempt to estimate the presence and strength of interactions between species from time-series data. Time-series data can be analyzed using more or less information about the temporal structure of the data. The most simple approach is to investigate the variance–covariance structure of the stationary distribution of population densities. For example, the presence of negative correlations between the abundances of species has been used to infer that the species are competitors. More information can be extracted from the data by analyzing it as a first-order autoregressive process. This involves fitting an autoregressive model such as Eq. (6.1), which uses information about abundances of species at time $t-1$ to predict changes in abundances between times $t-1$ and t. Below, we discuss these two approaches, which include zero time lags or one time lag in the analysis.

6.4.1 Zero time lags

An immediate conclusion from Figures 6.1 and 6.3 is that the covariance between the abundances of species may be a misleading measure of interactions between them. The two figures show simulations of the same model system, but the covariance between the two competitors is positive in Figure 6.2 and negative in Figure 6.3.

Negative covariance in the abundances of two species could indicate competition, but also depends on environmental correlation. Two species with similar diets that compete strongly may also show positively correlated responses to the environment due to their ecological similarity. Thus, the potential for negative covariance in abundances caused by competition is opposed by positive covariances caused by similar responses to environmental disturbances. The situation becomes muddier still when investigating more than two species. If the abundance of a strong competitor negatively covaries with the abundances of two other species, these two inferior competitors may covary positively in the presence of the strong competitor, thereby obscuring any competition between them. In short, the variance–covariance structure of the stationary distribution alone does not give much information about species interactions.

These comments apply not only to time-series data, but also attempts to infer species interactions from species abundances measured simultaneously at multiple sites. This has been done by regressing the abundance of a focal population on the abundances of other populations in the same habitat (Schoener 1974; Crowell and Pimm 1976). This regression technique, most often used to assess competition or mutualism, has been questioned by several authors both on theoretical and empirical grounds (Bender et al. 1984; Rosenzweig et al. 1985; Abramsky et al. 1986; Morris 1989; Brown and Heske 1990; Pfister 1995; Garrett and Dixon 1997; Shenbrot and Krasnov 2002), but defended and refined by others (Hallet and Pimm 1979; Pimm 1985; Schoener 1985; Carnes and Slade 1988; Fox and Luo 1996). A basic assumption in the regression method is that at any point in time, the abundance of a focal population is negatively correlated with the abundances of its competitors and positively correlated with its mutualists. This is a direct result of equilibrium analysis of common competition models, such as the Lotka–Volterra model (Schoener 1974). Not only does this method ignore correlated responses to environmental disturbances, it also relies on each population being at equilibrium and hence ignores natural variability in species abundances.

6.4.2 First-order lags

When time-series data are available, it makes sense to use the information available for changes in population abundances through time (Pfister 1995). For populations without a significant age- or stage-structure in environments with no autocorrelation the first-order autoregressive model in Eq. (6.1) may be a reasonable description of population dynamics. This can be fit directly to multispecies time-series data. The existence of measured environmental drivers (e.g. temperature) and measurement error can be incorporated into the statistical model. Environmental correlation in the disturbances affecting species present no problems unless environmental correlations create such strong correlations between the abundances of species that estimation of the autoregression coefficients becomes confounded by multi-colinearity.

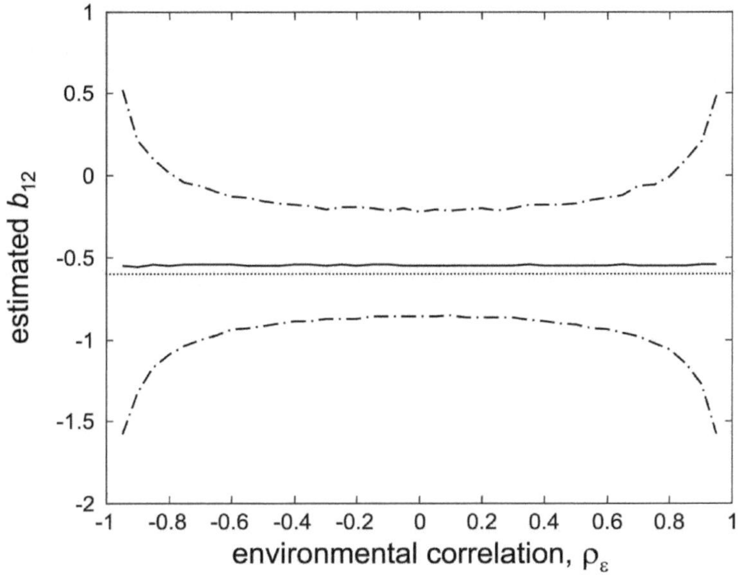

Figure 6.4. The estimated system matrix element b_{12} of a linear model (Eq. 6.1) fit to two parallel time series 30 time-steps long. The time series were generated with a system matrix $\mathbf{B} = \begin{pmatrix} 0.1 & -0.6 \\ -0.6 & 0.1 \end{pmatrix}$. Consequently the true value of b_{12} was -0.6. The correlation between the environmental fluctuations of the two (competing) populations varied between -0.95 and $+0.95$ (x-axis), and for each correlation value 1,000 time series were generated. From each simulated pair of time series, a new system matrix was estimated using the least square fit. Here the mean (solid line) and 95% confidence interval (dot-dashed lines) of the estimated b_{12} are depicted. The true value is also indicated (dotted line).

In Figure 6.4, we demonstrate that the estimate of an interaction coefficient is on average unaffected by environmental correlation (solid line), but that estimation uncertainty can become large in heavily (positively or negatively) correlated environments (the dash-dotted lines indicate the 95% confidence interval).

6.5 FOOD WEB DYNAMICS IN AUTOCORRELATED ENVIRONMENTS

Environmental autocorrelation can, like correlation, change the qualitative character of population dynamics. In this section, we give a brief review of the impact of environmental autocorrelation on food web dynamics. Next, we discuss implications for time-series analysis.

Following Ripa and Ives (2003) we extend the general autoregressive model (Eq. 6.1) to a model allowing for temporal structure in the environment:

$$\mathbf{x}(t) = \mathbf{B}\mathbf{x}(t-1) + \varphi(t) \tag{6.5a}$$

$$\varphi(t) = \Gamma\varphi(t-1) + \varepsilon(t). \tag{6.5b}$$

Here, **B** and **x**(t) are defined as before. The vector $\boldsymbol{\varphi}(t)$ contains the species-specific environmental fluctuations, and the matrix Γ describes how the present environment, $\boldsymbol{\varphi}(t)$, relates to the environment in the last time step, $\boldsymbol{\varphi}(t-1)$. The environmental process (Eq. 6.5b) includes the random vector $\boldsymbol{\varepsilon}(t)$, which is assumed to be normally distributed with covariance matrix Σ_ε and without temporal autocorrelation. The environmental dynamics described in Eq. 6.5b are structurally equivalent to the *population* dynamics of Eq. (6.1). Below, we will often make the simplifying assumption that Γ is diagonal, which means that the future environment of one population does not depend on the past environments of other populations, although they may still be correlated due to nonzero off-diagonal elements of Σ_ε. This assumption assures that all environmental processes are first-order, autoregressive processes (AR(1), Box and Jenkins 1970).

Again, the stochastic dynamics of the food web are most easily understood in terms of the eigenvectors and eigenvalues of the system matrix **B**. In the general model (Eqs. (6.5a,b), above) each population i has its own fluctuating environment φ_i with an autocorrelation given by the diagonal element γ_{ii} of the matrix Γ (assuming Γ is diagonal). Since each eigenvector is composed of several populations, the dynamics along each eigenvector will be subject to environmental disturbances of several different levels of autocorrelation. However, it can be argued that the effects of different autocorrelations are in principle additive, such that it is sufficient to treat them separately (Ripa and Ives 2003). In other words, by knowing the system's response to a single, arbitrary, level of autocorrelation it is straightforward to infer the response to several, species-specific, autocorrelated environments by simply adding the effects on for instance population variance. Consequently, we will here assume that the transformed environmental disturbances along the eigenvector i follow the first-order dynamics:

$$\omega_i(t) = \gamma\,\omega_i(t-1) + \eta_i(t) \tag{6.6}$$

where γ represents the environmental autocorrelation and $\eta_i(t)$ are temporally uncorrelated random variables with mean zero. The variance of the corresponding eigenvector component, $y_i(t)$, becomes (Roughgarden 1975a; Ripa and Heino 1999):

$$\mathrm{V}(y_{i,t}) = \frac{1+\lambda_i\gamma}{\left(1-\lambda_i^2\right)(1-\lambda_i\gamma)}\mathrm{V}(\omega_{i,t}). \tag{6.7}$$

According to Eq. (6.7) there is a matching, or *resonance*, between population dynamics and environmental dynamics such that the variance of a component will be high when the eigenvalue λ_i and the autocorrelation γ have the same sign, but lower when the signs are opposite (cf. Ives 1995). If the overall environment of a food web is positively autocorrelated, fluctuations along the eigenvectors with positive eigenvalues will be more pronounced than those along eigenvectors with

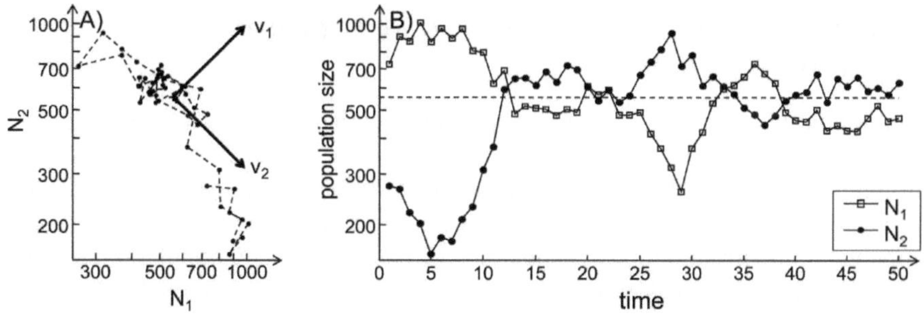

Figure 6.5. Simulation of the same competition model as in Figure 6.1, only with positively autocorrelated environments of the two competitors: $\varphi_{i,t+1} = \gamma\varphi_{i,t} + \varepsilon_{i,t}$, where $\varphi_{i,t}$ is the environment of population i (i = 1,2) at time t and $\varepsilon_{i,t}$ (i = 1,2) are serially uncorrelated, normally distributed variables with mean zero and variance $0.15^2(1 - \gamma^2)$, which gives the respective environments $\varphi_{i,t}$ a variance 0.15^2. The autocorrelation parameter γ is set to 0.8. (A) Phase portrait. (B) Time series of the same simulation.

negative eigenvalues, given the eigenvalues are of similar magnitude. Likewise, correlations between population densities will be determined by the directions of the eigenvectors with positive eigenvalues, which offers a link from environmental autocorrelation to population correlation.

In Figure 6.5 we show the effects of environmental autocorrelation on the simple competition model used above. A positively autocorrelated environment renders amplified dynamics along the eigenvector with a positive eigenvalue ($\mathbf{v}_2$) but weakens the dynamics along the other eigenvector ($\mathbf{v}_1$), which has a negative eigenvalue (Figure 6.5A). The positively autocorrelated environment gives positively autocorrelated population dynamics (Figure 6.5B), which is hardly surprising. However, it also gives a strong negative correlation between population densities, which is less intuitive (Figure 6.5A).

So far, we have only described the effects of environmental autocorrelation on population dynamics governed by real eigenvalues. Predator–prey interactions usually generate complex eigenvalues of the system matrix, which in turn generate population cycles. It can be shown (Ripa and Ives 2003) that the effect of environmental autocorrelation on a population cycle depends on the cycle length and the strength of the autocorrelation. As a rule of thumb, a long cycle (with a period length longer than four time steps) will increase in amplitude in a positively autocorrelated environment, compared to an uncorrelated environment, whereas a short cycle will decrease in amplitude (Ripa and Ives 2003).

In summary, the total effect on a food web of environmental autocorrelation can be quite profound, amplifying some dynamical patterns and reducing others. As a result, species interactions, such as competition or predator–prey interactions, can become either more or less visible in the food web dynamics (cf. Ripa and Ives 2003).

6.6 INTERACTION ASSESSMENTS IN AUTOCORRELATED ENVIRONMENTS

We now continue our discussion on the problems with time-series analysis caused by a structured environment. The usage of direct population correlations to assess population interactions is afflicted with serious concerns, one of them being the effects of correlated environments (above). As we have also demonstrated (Figure 6.5), environmental *auto*correlation can change the correlation between population densities, which adds another source of error to the "zero-lag" assessment method. A standard time-series analysis is to a large extent immune to the effects of environmental correlation (cf. Figure 6.4), but remains to be seen how environmental autocorrelation affects interaction estimates. We will here in particular study possible biases in parameter estimates due to an autocorrelated environment.

6.6.1 First-order lags

What would the bias be if we fit a first-order model (Eq. 6.1) to a time series from a community with autocorrelated environmental fluctuations? In Appendix 6A, we derive the approximate bias of the first-order community matrix **B**, given time-series data from a community experiencing first-order lags in environmental disturbances as in Eq. (6.5). Letting $\hat{\mathbf{B}}$ denote the estimated community matrix by (inappropriately) fitting Eq. (6.1) to a system given by Eq. (6.5) with true community matrix **B** we get:

$$\mathrm{E}(\hat{\mathbf{B}}) \approx \mathbf{B} + \Gamma \mathbf{V}_{\varphi x} \mathbf{V}_x^{-1} \approx \mathbf{B} + \Gamma \mathbf{V}_{\varphi} \mathbf{V}_x^{-1} \qquad (6.8)$$

here, $\mathbf{V}_{\varphi x}$ is a matrix containing the covariances between the environmental processes and the population densities, and $\mathbf{V}_x$ and $\mathbf{V}_{\varphi}$ are the covariance matrices of the population densities and environmental disturbances, respectively. The first approximation is asymptotic: it becomes an equality in the limit when the length of the time-series approaches infinity. Tentative numerical experiments (not presented) suggest it is a good approximation as long as the time series for separate species are not too positively autocorrelated. If the autocorrelation function of the time series remains significantly positive at lags close to or exceeding half the length of the time series, the asymptotic approximation tends to be inaccurate. The second approximation is a linearization in Γ, excluding all higher-order terms. Numerical experiments with diagonal Γ matrices (not presented) suggest that the second approximation is accurate as long as the environmental autocorrelations (i.e. the diagonal elements of Γ) are below 0.5 in magnitude. In general, the second approximation should be accurate as long as the eigenvalues of Γ are all small in magnitude. Neither of the approximations in Eq. (6.5) make use of the assumption that Γ is diagonal.

Equation 6.8 implies that if the population fluctuations are high compared to the environmental variability (i.e. $\mathbf{V}_x$ is large compared to $\mathbf{V}_{\varphi}$), then the bias in the estimated community matrix is small. This would be the case when the eigenvalues of **B** are large, because large eigenvalues of the community matrix yield high amplitude fluctuations even for moderate environmental disturbances. However, population variability $\mathbf{V}_x$ also depends on the autocorrelation of the environment (as mentioned in the previous section). Moreover, environmental correlation between the responses of different species changes both $\mathbf{V}_\varphi$ and $\mathbf{V}_x$ such that the combined effects of correlation and autocorrelation are hard to disentangle.

From Eq. (6.8), the bias in the community matrix $\hat{\mathbf{B}}$ estimated by (inappropriately) fitting a first-order autoregressive model to data with environmental autocorrelation affects all elements of the community matrix, even if there is no cross-autocorrelation in the environment (i.e. even if Γ is diagonal). An autocorrelated environment of a particular species creates a bias in all elements of $\hat{\mathbf{B}}$ on the same row. In Figure 6.6, we graph the expected matrix element estimates of a linear symmetric competition model, where the autocorrelation of the first population is on the x-axis. There are four bundles of curves in Figure 6.6, one for each matrix element. Each bundle consists of three curves, two of which correspond to the two approximations in Eq. (6.8). The third curve in each bundle is the mean estimate

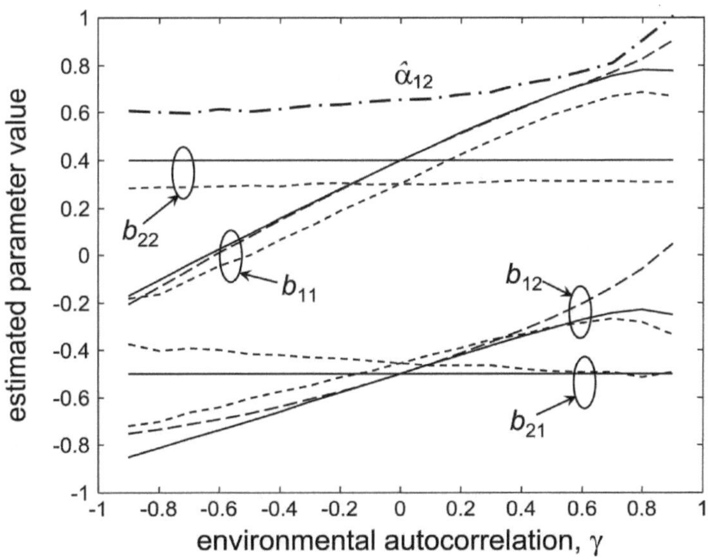

Figure 6.6. Matrix elements estimates from fitting a first-order model (Eq. 6.1) to time series of a second-order model (Eq. 6.5). The system matrix $\mathbf{B} = \begin{pmatrix} 0.4 & -0.5 \\ -0.5 & 0.4 \end{pmatrix}$. The environments of the two populations are independent, and normally distributed with zero mean and unit variance. The environment of the second population has no autocorrelation (Γ has only zeros on the second row), and the environmental autocorrelation of the first population varies along the x-axis. Three lines are drawn for each matrix element, as indicated. Solid line: first approximation in Eq. (6.8). Long dashed line: second approximation in Eq. (6.8). Short dashed lines: average estimates from the last 30 time steps of 1,000 simulations of the model with 100 initial time steps. Top, dot-dashed line: the standardized competition coefficient (from the averaged simulation estimates) from population 2 on population 1, as described in the text (Eq. 6.9).

from linear regression of 1,000 simulations of 30 time steps of the model. The figure gives a good illustration of the accuracy of the approximations in Eq. (6.8).

The autocorrelation of the environment of the second species in Figure 6.6 is zero, explaining the absence of bias in the estimates of b_{21} and b_{22}, except for the effect of a short time series (the difference between the simulation results and the two coinciding approximation curves). The estimates of the elements in the first row (b_{11} and b_{12}), however, show quite a large dependency on the environmental autocorrelation. As the environmental autocorrelation increases, the estimated intraspecific competition of population 1 decreases, as does the estimated competition from population 2 on population 1. If we had included environmental autocorrelation in the second species as well, there would be equivalent biases of the matrix elements of the second row. Consequently, even though competition is symmetric, environmental autocorrelation will make it appear asymmetric if the environmental autocorrelation of the two populations differs. Moreover, competition will seem less severe as the autocorrelation increases. This might seem to contradict our earlier result from a very similar model where environmental autocorrelation increased the negative correlation of competitors, thereby making the competition more "visible" (Figure 6.5). The contradiction is resolved, however, if we standardize the estimated interspecific competition with the *intra*specific competition (Levins 1968; Seifert and Seifert 1976; Laska and Wooton 1998):

$$\hat{\alpha}_{ij} = \frac{\hat{b}_{ij}}{\hat{b}_{ii} - 1}, \tag{6.9}$$

where the subtraction by one in the denominator stems from the discrete-time formulation. This standardized competition coefficient is analogue to the parameter α in the discrete-time Lotka–Volterra model (legend of Figure 6.1). As the environmental autocorrelation increases in our example, so does the estimated standardized competition from population 2 on population 1 (dash-dotted line in Figure 6.6).

We have sought a more useful and explicit interpretation of the expressions in Eq. (6.8), but failed. We find no easy way of predicting the direction and magnitude of the bias in parameter estimates caused by environmental autocorrelation. Numerical experiments show that the bias may not even be a monotonic function of the environmental autocorrelation, but often has a maximum or minimum at an intermediate level of autocorrelation (which is the case for b_{11} and b_{12} in Figure. 6.6).

6.6.2 Second-order lags

Given the risks of fitting a first-order autoregressive model to time series generated with environmental autocorrelation, it might seem best to fit a second-order model. As we show below, a second-order autoregressive model can capture the full structure of the stochastic process with environmental autocorrelation given by Eq. (6.5). Unfortunately, this approach is not without a new set of ambiguities.

Roughgarden (1975a) showed that a first-order autoregressive process with an autocorrelated environment is equivalent to a second-order autoregressive process. It is easy to show that our model community (Eq. 6.5) is equivalent to a system with delayed interactions

$$\mathbf{x}(t) = (\mathbf{B} + \Gamma)\mathbf{x}(t-1) - \Gamma\mathbf{B}\mathbf{x}(t-2) + \varepsilon(t). \tag{6.10}$$

Unless the environmental processes are known (i.e. $\varepsilon(t)$ or $\varphi(t)$), it is impossible to distinguish the system given by Eq. (6.10) from that in Eq. (6.5). Thus, suppose an ecologist with a time series identifies autocorrelation in the residuals after fitting a first-order autoregressive model (Eq. 6.1). If environmental autocorrelation were ignored, the response to autocorrelated residuals would be to fit a second-order autoregressive model,

$$\mathbf{x}(t) = \mathbf{B}_1\mathbf{x}(t-1) + \mathbf{B}_2\mathbf{x}(t-2) + \varepsilon(t) \tag{6.11}$$

The expected values of the estimated first- and second-order community matrices ($\mathbf{B}_1$ and $\mathbf{B}_2$) fit to a first-order autoregressive process with autocorrelated environmental disturbances (Eq. 6.5) are (cf. Eq. 6.10):

$$\mathrm{E}\left(\hat{\mathbf{B}}_1\right) = \mathbf{B} + \Gamma \tag{6.12a}$$

$$\mathrm{E}\left(\hat{\mathbf{B}}_2\right) = -\Gamma\mathbf{B} \tag{6.12b}$$

The estimate of first-order interactions, $\mathbf{B}_1$, are thus biased, and second-order interactions are falsely identified. If the assumption that Γ is a diagonal matrix is valid, the bias in the first-order interactions would be confined to the diagonal elements of $\hat{\mathbf{B}}_1$. Hence, the first-order interactions between populations could still be correctly estimated, but in the presence of positive (negative) environmental autocorrelation, there is bias towards less (more) severe intraspecific competition.

Since the bias of first-order interactions obtained from the fitted second-order model are confined to the diagonal elements of the first-order community matrix and have a simple relationship to the magnitude of environmental autocorrelation, fitting a second-order autoregressive model has advantages over fitting a first-order model when there is environmental autocorrelation. Nonetheless, environmental autocorrelation and true second-order species interactions remain confounded. The only way to get around this problem is to measure at least the most important environmental factors and include them explicitly in the model, rather than simply treat them as unmeasured environmental noise summarized in $\varepsilon(t)$. This is demonstrated in Figure 6.7, where one of two competitors is in part subject to an autocorrelated environmental factor. To be precise, the environment of population 1 is a sum of two environmental processes with equal variance, one of which is autocorrelated and the other is uncorrelated white noise. Population 2 is subjected to totally uncorrelated environmental disturbances with a variance equal to that of the total environment of population 1. Thus, only a fourth of the environmental variance

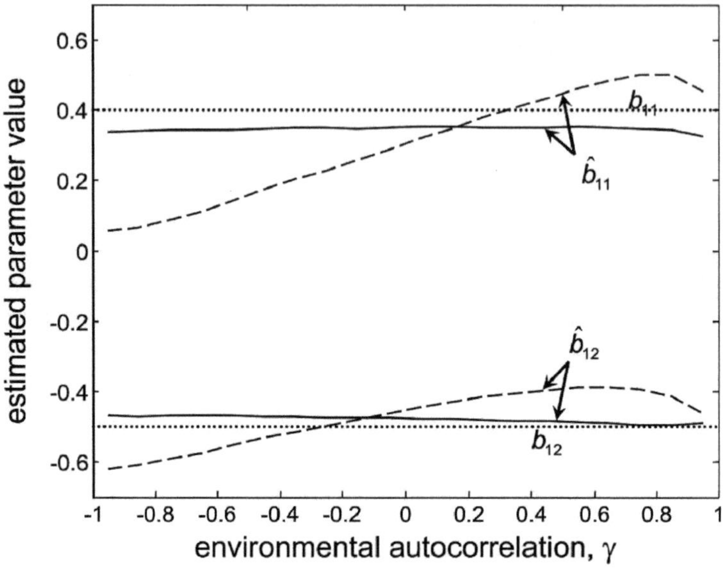

Figure 6.7. Estimated matrix elements of the first row of a two-by-two system matrix (b_{11} and b_{12}) of a linear food web model (Eq. 6.1). The environment of the first population, $\varphi_{1,t}$, is a sum of two components with equal variance (0.5). The first component is an AR(1) process according to Eq. (6.3) with autocorrelation parameter γ (x-axis). The second component of the environment of the first population is serially uncorrelated, i.e. white noise, and normally distributed. The stochastic environment of population two has no autocorrelation and is normally distributed with mean zero and unit variance. For each value of γ, 10,000 time series of length 30 are generated (with 100 initial time steps), and for each simulation a new system matrix is estimated using least squares. The resulting mean parameter values depended on whether the autocorrelated part of the first population's environment was used as a covariate in the analysis (solid lines) or not (dashed lines). The dotted lines indicate the true parameter values. The system matrix is the same as in Figure 6.6.

disturbing the community is autocorrelated. Yet it creates severe biases in parameter estimates (Figure 6.7, dashed lines). However, if the autocorrelated environmental factor affecting population 1 is included as a covariate in the regression analysis, the bias disappears (Figure 6.7, solid lines), except for the effect of a short time series.

6.7 DISCUSSION

We have here addressed some of the problems of assessing population interactions from population abundance data. First of all, we conclude that simple between-population correlation, or equivalent regression analysis, is not a good measure of interaction strength or even type. The possible, and likely, correlation between populations' responses to environmental fluctuations will alter the correlation between population sizes, making correlation analysis an unreliable tool for interaction assessment. As an example, two competitors can, depending on the environmental correlation alone, exhibit zero, positive, or negative correlation between their respective population sizes (Figures 6.1–6.3). This argument holds even if population sizes are sampled at different points in space rather than in time. There are additional, severe, problems afflicted with the correlation approach, often called the "static regression" technique or something similar, pointed out by several authors (Bender et al. 1984; Rosenzweig et al. 1985; Abramsky et al. 1986; Morris 1989; Pfister 1995; Garrett and Dixon 1997; Shenbrot and Krasnov 2002).

The problem with environmental correlation discussed above can be circumvented if the temporal structure of population data is taken into account. Fitting an appropriate ecological model to multispecies time series will (ideally) correctly identify and quantify population interactions, save for the ubiquitous estimation uncertainties, with relatively minor problems caused by environmental correlation (Figure 6.4). At least in the case of linear population dynamics, correlation causes no estimation bias, only an increased uncertainty in heavily correlated environments (Figure 6.4).

Multivariate time-series analysis has lately become increasingly popular among ecologists but has been used occasionally since at least a couple of decades (e.g. Cohen and Stone 1987, Elkinton et al. 1996, Ives et al. 1999). Given that the appropriate data can be acquired it offers a convenient method of interaction assessment that does not perturb the system in focus. Another major advantage is the direct relationship between model and data. Food web models predict population growth rates as functions of populations densities, both of which are directly available from time-series data.

We find the time-series analysis approach to work well as long as environmental fluctuations are uncorrelated over time. In an autocorrelated environment, however, estimated parameter values will be biased (Eqs. (6.8) and (6.12), Figures 6.5 and 6.6). A food web disturbed by autocorrelated environmental fluctuations will behave as if time-delayed interactions were present. If this is ignored and a first-order model is fit to data, all interaction coefficients will be biased (Eq. 6.8). The bias is the weakest if the food web dynamics are close to unstable, i.e. if population variability is large compared to environmental variability. As an example, the interactions in a strong predator–prey cycle could be measured with little bias, despite an autocorrelated environment. If the apparent time-delay is taken into account some parameter estimates might be unbiased, but on the other hand false delayed effects will be identified (Eq. 6.12). The bias and false delayed interactions disappear if the autocorrelated environmental factor is included as a covariate in the time-series analysis, which underlines the importance of environmental data in the analysis of ecological systems. Although it is impossible to measure every detail of the abiotic and biotic environment of a population or a community, at least some key environmental factors should be measured or collected from other databases, which might reduce the error and bias of the interaction assessments considerably (cf. Figure 6.7). Having said that, we do acknowledge the difficulty in choosing exactly which components of a multidimensional fluctuating environment should be included as covariates in a time-series analysis (cf. Ranta et al., Chapter 5, this volume). Even a single environmental factor, such as temperature, offers a multitude of possible input data: yearly, monthly, or weekly means, minimal/maximal values, rate of change, etc. Often, prior biological knowledge is insufficient and biologists are left with educated guesses and/or statistical model selection methods. Model selection is uncertain, however, if the number of data sources to choose from is too large (due to spurious correlations) or too small (such that critical factors are left out). Based on our results, we can at least offer one piece of advice: environmental factors with strong temporal autocorrelation should be given extra attention since they give biased interaction estimates if incorrectly ignored (Eq. (6.8), Figure 6.7).

Our analysis and numerical demonstrations are based on linear food web models. Interactions are for sure non-linear and most certainly often non-independent (Abrams 2002; Peacor and Werner 2004), but we find no reason to believe that non-linearities or dependencies between interactions should weaken our case. The correlation and autocorrelation of environmental fluctuations may cause severe problems for interaction assessments and it is imperative for ecologists to include environmental drivers in any analysis of ecological time series.

6.8 ACKNOWLEDGMENTS

First of all, we thank Kevin McCann and David Vasseur for the opportunity to publish in this volume. Secondly, we would like to thank Marm Kilpatrick, Jennifer Klug, Bea Beisner, Niclas Jonzén, Per Lundberg, and Esa Ranta for valuable discussions and comments on earlier drafts of this manuscript. This project was funded by grants from the National Science Foundation (USA). In addition, Jörgen Ripa was financially supported by the Swedish Research Council and the Royal Physiographic Society in Lund (Sweden).

6.9 LITERATURE CITED

Abramsky, Z., M.A. Bowers, and M.L. Rosenzweig. 1986. Detecting interspecific competition in the field: testing the regression method. *Oikos* 47: 199–204.

Abrams, P.A. 2002. Describing and quantifying interspecific interactions: a commentary on recent approaches. *Oikos* 94: 209–218.

Bender, E.A., T.J. Case, and M.E. Gilpin. 1984. Perturbation experiments in community ecology: theory and practice. *Ecology* 65: 1–13.

Berryman, A.A. 1991. Can economic forces cause ecological chaos? The case of the Northern California Dungeness crab fishery. *Oikos* 62: 106–109.

Box, G.E.P. and G.M. Jenkins. 1970. *Time series analysis: forecasting and control*. Holden-Day, San Francisco, CA.

Brown, J.H. and E.J. Heske. 1990. Temporal changes in a Chihuahuan desert rodent community. *Oikos* 59: 290–302.

Carnes, B.A. and N.A. Slade. 1988. The use of regression for detecting competition with multicollinear data. *Ecology* 69: 1266–1274.

Caswell, H. and J.E. Cohen. 1995. Red, white and blue: environmental variance spectra and coexistence in metapopulations. *Journal of Theoretical Biology* 176: 301–316.

Cohen, Y. and J.N. Stone. 1987. Multivariate time series analysis of the Canadian fisheries system in Lake Superior. *Canadian Journal of Fisheries and Aquatic Sciences* 44: 171–181.

Crowell, K.L. and S.L. Pimm. 1976. Competition and niche shifts of mice introduced onto small islands. *Oikos* 27: 251–258.

Cuddington, K.M. and P. Yodzis. 1999. Black noise and population persistence. *Proceedings of the Royal Society Biological Sciences Series B* 266: 969–973.

Elkinton, J.S., W.M. Healy, J.P. Buonaccorsi, G.H. Boettner, A.M. Hazzard, H.R. Smith, and A.M. Liebhold. 1996. Interactions among gypsy moths, white-footed mice, and acorns. *Ecology* 77: 2332–2342.

Elton, C.S. 1924. Periodic fluctuations in the numbers of animals: their causes and effects. *British Journal of Experimental Biology* 2: 119–163.

Elton, C. and M. Nicholson. 1942. The ten year cycle in numbers of the lynx in Canada. *Journal of Animal Ecology* 11: 215–244.

Fox, B.J. and J. Luo. 1996. Estimating competition coefficients from census data: a re-examination of the regression technique. *Oikos* 77: 291–300.

Garrett, K.A. and P.M. Dixon. 1997. Environmental pseudointeraction: the effects of ignoring the scale of environmental heterogeneity in competition studies. *Theoretical Population Biology* 51: 37–48.

Griebeler, E.M. and E. Gottschalk. 2000. The influence of temperature model assumptions on the prognosis accuracy of extinction risk. *Ecological Modelling* 134: 343–356.

Hallet, J. and S.L. Pimm. 1979. Direct estimation of competition. *The American Naturalist* 113: 593–600.

Halley, J.M. and V.E. Kunin. 1998. Extinction risk and the 1/f family of noise models. *Theoretical Population Biology* 56: 215–230.

Heino, M. 1998. Noise colour, synchrony and extinctions in spatially structured populations. *Oikos* 83: 368–375.

Ives, A.R. 1995a. Measuring resilience in stochastic systems. *Ecological Monographs* 65: 217–233.

Ives, A.R., S.R. Carpenter, and B. Dennis. 1999. Community interaction webs and zooplankton responses to planktivory manipulations. *Ecology* 80: 1405–1421.

Jonzén, N., P. Lundberg, E. Ranta, and V. Kaitala. 2002. The irreducable uncertainty of the demography–environment interaction in ecology. *Proceedings of the Royal Society Biological Sciences Series B* 269: 221–225.

Kaitala, V., J. Ylikarjula, E. Ranta, and Lundberg, P. 1997. Population dynamics and the colour of the environmental noise. *Proceedings of the Royal Society Biological Sciences Series B* 264: 943–948.

Laska, M.M. and J.T. Wooton. 1998. Theoretical concepts and empirical approaches to measuring interaction strength. *Ecology* 79: 461–476.

Lehman, C.L. and D. Tilman. 2000. Biodiversity, stability, and productivity in competitive communities. *The American Naturalist* 156: 534–552.

Levins, S.A. 1968. Evolution in changing environments. Princeton University Press, Princeton, NJ.

Morris, D.W. 1989. Habitat-dependent estimates of competitive interaction. *Oikos* 55: 111–120.

Peacor, S.D. and E.E. Werner. 2004. How dependent are species–pair interaction strengths on other species in the food web? *Ecology* 85: 2754–2763.

Petchey, O.L., A. Gonzalez, and H.B. Wilson. 1997. Effects on population persistence: the interaction between environmental noise colour, intraspecific competition and space. *Proceedings of the Royal Society Biological Sciences Series B* 264: 1841–1847.

Pfister, C.A. 1995. Estimating competition coefficients from census data: a test with field manipulations of tidepool fishes. *The American Naturalist* 146: 271–291.

Pimm, S.L. 1985. Estimating competition coefficients from census data. *Oecologia (Berl.)* 67: 588–590.

Pimm, S.L. and A. Redfearn. 1988. The variability of animal populations. *Nature* 334: 613–614.

Reinsel, G.C. 1997. *Elements of multivariate time series analysis*, 2nd ed. Springer, New York.

Ripa, J. and M. Heino, M. 1999. Linear analysis solves two puzzles in population dynamics: the route to extinction and extinction in coloured environments. *Ecology Letters* 2: 219–222.

Ripa, J. and A.R. Ives. 2003. Food web dynamics in correlated and autocorrelated environments. *Theoretical Population Biology* 64: 369–384.

Ripa, J. and P. Lundberg. 1996. Noise colour and the risk of population extinctions. *Proceedings of the Royal Society Biological Sciences Series B* 263: 1751–1753.

Ripa, J., P. Lundberg, and V. Kaitala. 1998. A general theory of environmental noise in ecological food webs. *The American Naturalist* 151: 256–263.
Rosenzweig, M.L., Z. Abramsky, B. Kotler, and W. Mitchell. 1985. Can interaction coefficients be determined from census data? *Oecologia (Berl.)* 66: 194–198.
Roughgarden, J. 1975a. A simple model for population dynamics in stochastic environments. *The American Naturalist* 109: 713–736.
Roughgarden, J. 1975b. Population dynamics in a stochastic environment: spectral theory for the linearized N-species Lotka–Volterra competition equations. *Theoretical Population Biology* 7: 1–12.
Royama, T. 1981. Fundamental concepts and methodology for the analysis of animal population dynamics, with particular reference to univoltine species. *Ecological Monographs* 51: 473–493.
Schoener, T.W. 1974. Competition and the form of habitat shift. *Theoretical Population Biology* 6: 265–307.
Schoener, T.W. 1985. On the degree of consistency expected when different methods are used to estimate competition coefficients from census data. *Oecologia (Berl.)* 67: 591–592.
Seifert, R.P. and F.H. Seifert. 1976. A community matrix analysis of Heliconia insect communities. *The American Naturalist* 110: 461–483.
Shenbrot, G. and B. Krasnov. 2002. Can interaction coefficients be determined from census data? Testing two estimation methods with Negev Desert rodents. *Oikos* 99: 47–58.
Steele, J.H. 1985. A comparison of terrestrial and marine ecological systems. *Nature* 313: 355–358.
Strebel, D.E. 1985. Environmental fluctuations and extinction – single species. *Theoretical Population Biology* 27: 1–26.
Walters, C.J. 1990. A partial bias correction factor for stock-recruitment parameter estimation in presence of autocorrelated environmental effects. *Canadian Journal of Fisheries and Aquatic Sciences* 47: 516–519.
Williams, D.W. and A.M. Liebhold. 1995. Detection of delayed density dependence: effects of autocorrelation in an exogenous factor. *Ecology* 76: 1005–1008.

APPENDIX 6A FIRST-ORDER SYSTEM MATRIX BIAS IN AUTOCORRELATED ENVIRONMENTS

Consider the stochastic system in Eq. (6.5). Given time-series data of $\mathbf{x}(t)$, a first-order regression of the $\mathbf{x}(t)$ process on itself gives

$$\hat{\mathbf{B}} = \hat{\mathbf{C}}\hat{\mathbf{V}}_x^{-1}, \tag{6A.1}$$

where $\hat{\mathbf{C}}$ is the estimation of $\mathrm{cov}(\mathbf{x}_t, \mathbf{x}_{t-1})$ and $\hat{\mathbf{V}}_x$ is the estimated covariance matrix of the population densities ($\mathrm{cov}(\mathbf{x}_t, \mathbf{x}_t)$). As the length of the time series goes to infinity, the expected error in the estimates goes to zero. Thus, an approximation of the expectation of the system matrix is

$$\mathrm{E}(\hat{\mathbf{B}}) \approx \mathbf{C}\mathbf{V}_x^{-1}, \tag{6A.2}$$

where $\mathbf{C} = \mathrm{cov}(\mathbf{x}_t, \mathbf{x}_{t-1})$ and $\mathbf{V}_x = \mathrm{cov}(\mathbf{x}_t, \mathbf{x}_t)$. Multiplying Eq. (6.5a) by $\mathbf{x}^T(t-1)$ from the right and taking the expected values gives

$$\mathbf{C} = \mathbf{B}\mathbf{V}_x + \mathrm{cov}(\boldsymbol{\varphi}(t), \mathbf{x}(t-1)) = \mathbf{B}\mathbf{V}_x + \mathrm{cov}(\boldsymbol{\Gamma}\varphi(t-1) + \varepsilon(t), \mathbf{x}(t-1)) = \mathbf{B}\mathbf{V}_x + \boldsymbol{\Gamma}\mathbf{V}_{\varphi x}, \tag{6A.3}$$

where $\mathbf{V}_{\varphi x} = \mathrm{cov}(\boldsymbol{\varphi}(t), \mathbf{x}(t))$. Inserting Eq. (6A.3) into Eq. (6A.2) gives the asymptotic approximation

$$\mathrm{E}(\hat{\mathbf{B}}) \approx \mathbf{B} + \Gamma\mathbf{V}_{\varphi x}\mathbf{V}_x^{-1} \tag{6A.4}$$

From Eq. (6.5),

$$\mathbf{V}_{\varphi x} = \mathrm{cov}(\varphi(t), \mathbf{x}(t)) = \mathrm{cov}(\varphi(t), \mathbf{Bx}(t-1) + \varphi(t)) = \mathrm{cov}(\Gamma\varphi(t-1) + \varepsilon(t), \mathbf{Bx}(t-1)) + \mathbf{V}_{\varphi} = \Gamma\mathbf{V}_{\varphi\mathrm{x}}\mathbf{B}^{\mathrm{T}} + \mathbf{V}_{\varphi} \qquad (6\mathrm{A}.5)$$

Repeatedly inserting Eq. (6A.5) into itself yields

$$\mathbf{V}_{\varphi x} = \mathbf{V}_{\varphi} + \Gamma\mathbf{V}_{\varphi}\mathbf{B}^{\mathrm{T}} + \Gamma^2\mathbf{V}_{\varphi}\mathbf{B}^{2\mathrm{T}} + \cdots = \sum_{k=0}^{\infty}\Gamma^k\mathbf{V}_{\varphi}\mathbf{B}^{k\mathrm{T}} \qquad (6\mathrm{A}.6)$$

Inserting Eq. (6A.6) into Eq. (6A.4) and excluding all higher-order terms gives

$$\mathrm{E}(\hat{\mathbf{B}}) \approx \mathbf{B} + \Gamma\mathbf{V}_{\varphi}\mathbf{V}_x^{-1} \qquad (6\mathrm{A}.7)$$

CHAPTER 7

SPECIALIST–GENERALIST COMPETITION IN VARIABLE ENVIRONMENTS; THE CONSEQUENCES OF COMPETITION BETWEEN RESOURCES

PETER A. ABRAMS

Department of Ecology and Evolutionary Biology, University of Toronto, 25 Harbord St. Toronto, Ontario M5S 3G5, Canada
Phone: 416-978-1014 Fax: 416-978-5878 E-mail: abrams@zoo.utoronto.ca

7.1 ABSTRACT

Consumer-resource models have long been employed to help understand competition between consumers, but the resources have usually been assumed to be independent of each other. The impact of competition between resources is explored for a case in which two specialists and one generalist compete for two nutritionally substitutable resources. Coexistence of all three consumers requires that the relative abundance of the two resources fluctuates over time. The conditions allowing coexistence of all three consumer types is generally broadened considerably by the presence of competition between resources. In addition, competition between resources usually increases the density of the generalist consumer relative to the specialists. These

David A. Vasseur and Kevin S. McCann, The Impact of Environmental Variability on Ecological Systems, 133–157. © 2007 *Springer*.

results are due to the fact that the generalist persists by "consuming" negative covariance between resources, and competition between the resources increases the negative covariance in the absence of the generalist. The densities of the competing consumers often respond in counterintuitive ways to mortality imposed on one of the consumers. Frequently, the mean population size of one consumer is very insensitive to its increasing mortality over a broad range of mortality rates, but abruptly goes extinct when mortality surpasses a threshold value.

Keywords: competition, specialist, generalist, coexistence.

7.2 INTRODUCTION

During the 1970s several researchers discovered that sustained fluctuations in resource abundance could allow two or more species to coexist on a single resource (Stewart and Levin 1973; Koch 1974; Armstrong and McGehee 1976, 1980; Hsu et al. 1978; Levins 1979). Most experimental and theoretical work on exceptions to the "competitive exclusion principle" since then has focused on systems in which a single resource is the subject of competition. In a recent review, Chesson (2000) noted that very little was known about systems in which consumers competed for two or more variable resources. The few studies that have investigated competition for two or more resources (Huisman and Weissing 1999, 2001, 2002) have investigated a narrow range of models, and have assumed that the resources did not affect each other's growth rates. The present chapter was largely motivated by a desire to determine the consequences for competitive coexistence of more species than resources in nonstationary systems when: (i) there is more than a single limiting resource; and (ii) the different resources interact with each other. This chapter will be restricted to the simplest multi-resource model, namely two resources that are nutritionally substitutable for their consumers. It will focus on the impact of competition between the resource populations on the interaction of the consumers that are competing for those resources. The framework with two substitutable resources builds on other recent studies of competition, coexistence, and coevolution in this framework (Abrams 2006a,b,c). Investigating the effect of resource competition on the interactions of consumers that share those resources was motivated by a series of articles by John Vandermeer (1980, 2004), showing that resource competition could greatly alter conditions for the coexistence of consumers for models in which the number of consumers was assumed to be less than or equal to the number of resources.

The work described here builds upon previous 1- and 2-resource models of competition for resources having temporal fluctuations in abundance, so it is useful to begin by reviewing some of the conclusions reached by that work. Analysis of the conditions required for coexistence of two species on a single resource suggests that endogenous consumer–resource cycles are more likely to lead to coexistence than is environmental forcing of resource growth (Abrams 2004), and that coexistence over a broad range of parameter values requires large differences in the shapes of the functional responses of the two species (Abrams and Holt 2002). Here I concentrate on systems with consumers having similar functional response shapes, to determine whether the presence of two resources can allow three or more species to coexist by a fundamentally different mechanism than the difference in response shape that underlies most single-resource models.

Most of the previous studies of competition for a single fluctuating resource have reached the conclusion that it would require very precise balancing of parameters for more than two species to coexist on one resource purely via differences in the nonlinearity of their response. The exceptions to this generalization are studies that allow for extremely low densities during the course of cycles (Ebenhöh 1988; Litchman and Klausmeier 2001). Chesson (1994) concluded that it was unlikely that differences in the linearity of the functional responses of different species could result in coexistence of a large number of species on a single resource, and Abrams and Holt (2002) and Abrams (2004) both note that coexistence of more than two species only occurred over very restricted ranges of parameters in traditional consumer-resource models in which resource populations fluctuated either because of sinusoidal environmental forcing of resource growth or because of endogenously generated consumer–resource cycles. Thus, consumer-resource models having only a single resource lead one to believe that coexistence of one more consumer than the number of resources might be achieved easily, but coexistence of more than one additional species should be relatively rare. It is still unclear to what extent this conclusion extends to systems two or more resources, and this question will be a central focus of this chapter.

Another aspect of competition for variable resources that has received relatively little attention is the change in the impact of one consumer species on the average density of a competing consumer, given nonstationary resource populations. Abrams et al. (2003) examined the responses to mortality of one consumer in a system of two consumers competing for one resource, as described by Armstrong and McGehee (1980). They found that average densities of the two consumers responded in a highly nonlinear and often counterintuitive way to mortality imposed on one consumer. It was possible for an increase in the mortality of one of the species to increase the mean population densities of both, and there were many cases where a small change in mortality applied to one consumer species led to a discontinuous jump in the average density of the competitor. It is not known whether these features are likely to be restricted to the special case of two competitors for a single resource, or whether they might apply more broadly to systems in which species compete for fluctuating resources, or to systems that are "supersaturated" (Schippers et al. 2001) in the sense of having more consumers than resources.

The reason for extending work on supersaturated coexistence from single-resource systems to multiple-resource systems, is not simply that it has received little attention. The more important motivations are that natural systems invariably have more than a single resource, and that having two or more resources allows fundamentally different mechanisms to operate. This is illustrated by the work of Huisman and Weissing (1999, 2001, 2002), who examined the ability of more than three consumers to coexist on three abiotic resources. Huisman and Weissing modeled a situation in which all of the resources were essential to each consumer in the sense of Leon and Tumpson (1975). By having three resources, it was possible to set up a model in which there was an intransitive network of competitive abilities among three of the consumer species. Earlier work by May and Leonard (1975) and Coste et al. (1978) had shown that such intransitive hierarchies could generate cycles. In Huisman and Weissing's model, the cycles allowed still more species to coexist with the three that were driving the fluctuations in resource densities. However, other authors have argued that the parameter space allowing multispecies coexistence in this model is relatively narrow (Schippers et al. 2001), and the cycles required for coexistence only occur under the somewhat unlikely assumption that each consumer species is best at catching the resource it needs in intermediate amounts. One would expect that, if there were trade-offs between different minimum resource requirements, a species would evolve to become best at catching the resource that is likely to be most limiting to growth.

Aside from Huisman and Weissing's work, the only other ecological scenario in which coexistence of more consumers than resources has been studied in a general framework with two or more resources is the case of specialist–generalist competition for two nutritionally substitutable resources. Although a large number of ecological models have considered specialist–generalist competition, very few of these have examined situations under which the resources exhibit sustained fluctuations in density. Wilson and Yoshimura (1994) reviewed 43 previous models of evolution of resource use within a species, and none of these had included environmental variability. Wilson and Yoshimura (1994) were apparently the first to study this case, and they argued that coexistence of three species was a robust outcome that occurred over a wide range of parameter space provided that resources exhibited sustained fluctuations in abundance and that the consumers exhibited adaptive behavioral choice of resources. In their model, the resources were represented by a consumer-species-specific carrying capacity in each of two habitats, and the carrying capacity of each habitat exhibited independent temporal variation. Yoshimura's (1994) work was recently challenged by Egas et al. (2004), who suggested that using a different trade-off to determine the abilities of specialists and generalists to use resources greatly reduced the ability of species to coexist. In related evolutionary models, Kisdi (2002) and Parvinen and Egas (2004) failed to obtain coexistence of three species on two resources over a significant range of parameter space. The use of logistic growth models in all of these studies makes it difficult to determine which ones might be consistent with more detailed models in which resource dynamics and more individual-level traits are represented explicitly. The lack of explicit resources also makes it difficult to compare these results with the earlier work on 2-consumer-1-resource systems reviewed above, or with Huisman and Weissing's work on competition for two or more essential resources.

There are a number of reasons why it would be desirable to know more about the possibility for and nature of competitive coexistence of generalists and specialists in systems with environmental variability when the number of species exceeds the number of resources. Competition, coexistence, and evolution in specialist–generalist systems have all been topics of long-standing interest (Futuyma and Moreno 1988; reviewed in Wilson and Yoshimura 1994). A wide variety of plants are consumed by a range of specialist and generalist herbivores (e.g. Lawton et al. 1993). The fact that most resources are seasonal implies that environmental forcing of resource growth should be common. The review of long-term population studies by Kendall et al. (1998) shows that a substantial fraction of the populations undergo regular cycles that appear to be due to consumer–resource interactions. Thus, conditions that might be conducive to coexistence seem to be common.

The following section of this chapter begins by summarizing recent work (Abrams 2006a) that used a consumer–resource model to explore the potential coexistence of specialists and generalists on a pair of resources. That article showed that such coexistence could occur under a relatively wide range of circumstances when there was temporal variation in resource populations. The first part of this chapter extends this work to examine the impact of interactions between the two resources for competitive coexistence. The second part of the chapter looks at the population dynamical phenomena that can occur when two specialist and one generalist consumer coexist as a consequence of fluctuating resource populations. This section explores the impact of mortality rates of specialist or generalist on the population size of the two other consumer species. In many cases, these impacts are quite different from what would be expected for an analogous system in which two generalists compete for two resources, either under constant or variable conditions. Both of these results suggest that many traditional assumptions about the nature of competition may need to be revised when trying to understand systems with coexisting specialists and generalists having nonstationary dynamics and sharing resources that are themselves competitors.

7.3 WHEN AND WHY ARE SPECIALISTS AND GENERALISTS ABLE TO COEXIST ON TWO RESOURCES?

Two models will be discussed here; they correspond closely to the models considered in Abrams (2006a). The first assumes that generalists have a fixed preference, so that their attack rates on each of the two resources remain constant. The second assumes that the generalist can alter its preference on a behavioral timescale, and that the two attack rates are subject to a trade-off relationship. The reason for including behavioral switching is because it is likely to be common in nature (Murdoch 1969) and it was claimed to be essential for specialist generalist coexistence by both Wilson and Yoshimura (1994) and Egas et al. (2004).

7.3.1 Model 1: Fixed preferences

Equations (7.1) describe competition between three consumer species for two resources in a single habitat where the resources are encountered in a fine-grained manner. The two specialist consumers have population sizes N_1 and N_2, while the generalist has a population N_g, and a preference that is determined by a parameter p. The two resources may compete with each other, and have population sizes R_1 and R_2. The population dynamics are given by the following equations:

$$\begin{aligned}
\frac{dR_1}{dt} &= I_1 + r_1 R_1 \left(1 - \frac{R_1 + \alpha_{12} R_2}{K_1}\right) - \frac{C_1 N_1 R_1}{1 + hC_1 R_1} \\
&\quad - \frac{p^n C_1 N_g R_1}{1 + h\,(p^n C_1 R_1 + (1-p)^n C_2 R_2)} \\
\frac{dR_2}{dt} &= I_2 + r_2 R_2 \left(1 - \frac{R_2 + \alpha_{21} R_1}{K_2}\right) - \frac{C_2 N_2 R_2}{1 + hC_2 R_2} \\
&\quad - \frac{(1-p)^n C_2 N_g R_2}{1 + h\,(p^n C_1 R_1 + (1-p)^n C_2 R_2)} \\
\frac{dN_1}{dt} &= N_1 \left(\frac{b_1 C_1 R_1}{1 + hC_1 R_1} - d_1\right) \qquad\qquad (7.1\text{a–e}) \\
\frac{dN_2}{dt} &= N_2 \left(\frac{b_2 C_2 R_2}{1 + hC_2 R_2} - d_2\right) \\
\frac{dN_g}{dt} &= N_g \left(\frac{b_g p^n C_1 R_1 + b_g (1-p)^n C_2 R_2}{1 + h\,(p^n C_1 R_1 + (1-p)^n C_2 R_2)} - d_g\right)
\end{aligned}$$

The resources have logistic growth with parameters r_i and K_i, and external immigration given by I_i. Immigration rates are assumed to be small, and reflect the fact that resources are generally more widely distributed than are their consumers (Abrams and Holt 2002). External immigration also prevents the unrealistic cycle amplitudes that occur in many models with homogeneous prey populations and type-2 predator responses. The two resources may also compete with each other, as described by the competition coefficients, α_{ij} measuring the per capita effect of resource species j on the per capita growth rate of species i, relative to the effect of i on its own growth. The consumer functional responses are described by Holling (1959) disk equations, and are characterized by maximum attack rates on each resource j, C_j, and a common handling time, h. Consumer species i has a conversion efficiency of b_i and a per capita death rate of d_i. The index of specialization on resource 1, denoted p, ranges from 0 (when only resource 2 is consumed) to 1 (when only resource 1 is consumed). The exponent, n, denotes the shape of the trade-off. If $n < 1$, the generalist has a convex trade-off implying that a perfect generalist ($p = 1/2$) has a greater attack rate than the mean of the two specialists. If $n > 1$, the trade-off is concave, and the generalists have lower attack rates than the mean of the two specialists.

In many calculations, n is set to 1, implying a linear trade-off. In most cases with a fixed value of p, the generalist will be assumed to be characterized by $p = 0.5$, and $C_1 = C_2$, indicating equal abilities to consume each resource.

When the equilibrium of Eqs. (7.1) is stable, the competitive exclusion principle, as defined by Armstrong and McGehee (1980) assures us that at most two consumer species may coexist. This can be two specialists, or the generalist and one specialist. If the generalist has a sufficiently low resource requirement for zero population growth, it is also possible that the two specialists in this system are excluded by the generalist. In situations where the generalist has a linear trade-off ($n = 1$), and shares the same maximum attack rates, C_i, conversion efficiency, b, and handling time, h, with the specialists, then the mortality rate of the generalist relative to that of the specialists determines when each of these outcomes occurs. When the mortality of the generalist is less than that of the two specialists (and the latter two have the same mortality) then the generalist excludes both specialists; when the generalist mortality is greater, the generalist is excluded. When one specialist has a higher mortality than the generalist, while the other has a lower mortality, then the outcome is frequently coexistence of the generalist with the more efficient specialist. (Exclusion of a particular species also occurs in stable systems when it has a disadvantage in any of the other fitness-related parameters, C_i, b, or h. I concentrate on mortality here because it is most easily related to resource requirements, which are the most general measure of competitive ability.)

Systems with endogenous fluctuations in two resources must have some asynchrony in those fluctuations to allow coexistence of three (or more) consumer species; if both resources have identical densities at all times, then generalists always experience the same resource environment as specialists, and will exclude the specialists, or be excluded by them. Asynchrony in the resource dynamics gives a generalist a lower variance in total resource intake rate than the specialists. Due to the concave nature of the Holling type-2 functional responses, Jensen's Inequality (Ruel and Ayres 1999) implies that the lower variance translates into a higher mean growth rate. This allows a generalist to invade a system consisting of two cycling specialists even when the generalist has a greater mortality rate, lower resource conversion efficiency, or lower average resource capture rate. The advantage due to averaging resource densities disappears if the generalist becomes sufficiently abundant to synchronize the cycles in the two resources. A less efficient generalist will lose its advantage before the two resources are exactly synchronized. This ensures that the generalist does not exclude the specialists, unless it has a sufficiently low mortality (or high conversion efficiency). Asynchrony can arise in several different ways. Abrams (2006a) shows that a difference in the growth parameters of the two resources produces asynchronous fluctuations. Differences between the two specialist consumers in the magnitude of their demographic rates, b and d, can also produce asynchronous resource cycles (Abrams 2006a). Finally, competition between the two resources, a factor not considered in Abrams (2006a), often produces asynchrony in the cycles of the two resources (Vandermeer 2004). This means that competition between the resources can potentially allow coexistence in the absence of any differences between the two 1-specialist-1-resource components of the system. Another consequence of competition between resources is that the range of other parameters producing consumer–resource cycles expands greatly. In the limiting case where both competition coefficients are unity, then cycles occur in a system with two specialists for all nonzero handling times and all consumer R^* values when the resource immigration rates are zero (Nakajima et al. 2004). A third, well-known consequence of competition between resources is that the interaction between different consumers of those resources can be changed from competitive to mutualistic (Vandermeer 1980). Thus, it is clearly important to determine how competition between resources affects the coexistence of their consumers.

Figure 7.1 shows the dynamics of all three consumers and the two resources for two systems that differ in the magnitude of the generalist's disadvantage in fitness parameters relative to the specialists. The two resources compete with both competition coefficients given by $\alpha = 0.25$. Because the attack rate and handling time parameters of the functional responses are identical for the three consumers, the ratio of mortality rate to conversion efficiency is monotonically related to the equilibrium resource density, often denoted R^*. (For a consumer of two resources, a single R^* for that consumer can be defined as the equilibrium resource density under the constraint that all resources have the same equilibrium density.) When the generalist has an R^* value only slightly greater than that of the specialist, the dynamics shown in Figures 7.1A,C result, with relative synchrony of the three consumers with each other and of the two resources with each other. When the generalist has a larger R^*, due to the larger death rate given in Figures 7.1B,D, the generalist has little impact on the resource dynamics, which exhibit almost anti-symmetric cycles. This reduced synchrony gives the generalist an advantage, which is what allows it to persist even when it has an R^* value that is almost 50% greater than that of the specialists. The example in Figure 7.1 has specialists that differ in the magnitudes of their demographic rates (species 2 has a higher conversion efficiency and death rate than species 1). This would have been enough to produce asynchrony in the resource cycles in the absence of competition between them, but it would not have given rise to such a wide range of mortalities (or R^* values) allowing coexistence. Robust coexistence under Eqs. (7.1) requires that the consumers have strongly saturating functional or numerical responses, so that the reduced variation in the generalist's intake rate provides enough of an advantage to allow it to increase in spite of a higher resource requirement. The ratio of handling time to searching time when resources are at their carrying capacity (given by the product ChK for a specialist) provides an index of how rapidly the functional response approaches saturation as resource densities increase. In practice, values of ChK must be approximately 4 or larger for coexistence over a wide range of generalist mortality rates (Abrams 2006a).

Equations (7.1) assume that there is only a single generalist species, and it is certainly possible that several different generalist species are available to enter the 2-specialist community. If the two generalists have identical parameters except that

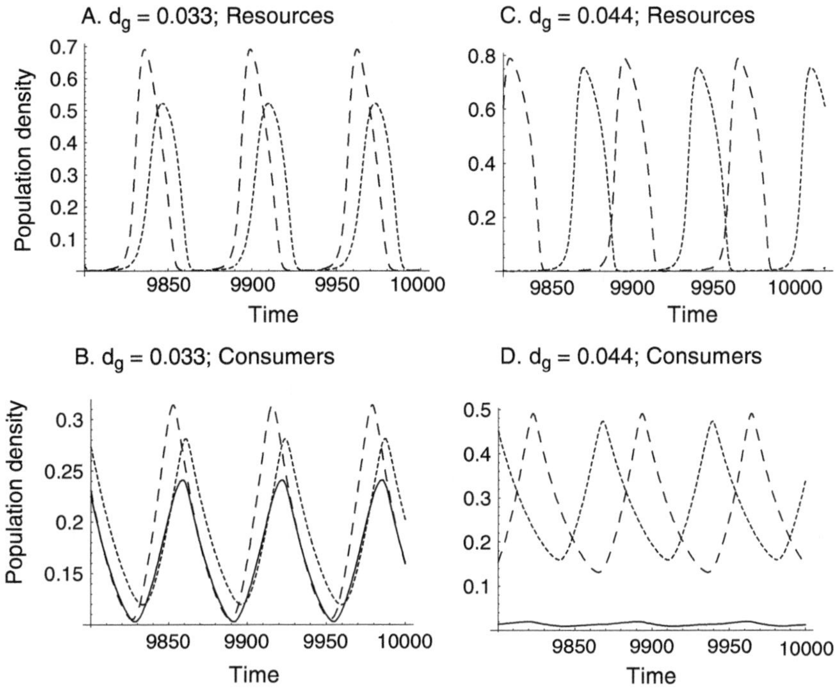

Figure 7.1. The dynamics of all three consumers and the two resources for two systems that differ in the magnitude of the generalist's efficiency disadvantage relative to the specialists. In the resource panels (a and b), resource 1 is the long-dashed line, while resource 2 is the short-dashed line. In the consumer panels (c and d), the generalist is given by the solid line, consumer 1 by the long-dashed line, and consumer 2 by the short-dashed line. In panels a and c, the generalist has a mortality rate of 0.033, while the two specialists have a corresponding value of 0.03. The modest phase shift in the cycles of the two resources represents a tension between the generalist consumption, which tends to synchronize the resources, and the competition between resources that tends to shift them to anti-synchrony (a 180 degree phase difference). In panels b and d, the generalist's mortality rate has been increased to near the maximum that allows it to persist, 0.044. The generalist's population size is low, and the phase difference between the resources is larger, since their dynamics have largely escaped the generalist's control. The parameter values in Eqs. (7.1) are: $r_1 = r_2 = K_1 = K_2 = 1$; $I_1 = I_2 = 0.001$; $b_1 = b_g = 0.1$; $b_2 = 0.12$; $d_1 = 0.03$; $d_2 = 0.036$; $C_1 = C_2 = 5$; $h = 1$; $n = 1$; $\alpha_{12} = \alpha_{21} = 0.25$.

one is more specialized on resource 1 (e.g. $p = 0.75$) and the other on resource 2 (e.g. $p = 0.25$), they will coexist with the specialists, provided the resources have both similar abundances and variance in their abundances when only the specialists are present. However, a very small ($\ll$10%) increase in mortality of one of these species is sufficient to drive it extinct. Moreover, extending Eqs. (7.1) so that the choice parameter p can evolve along the trade-off relationship results in one generalist lineage excluding all of the others, if the system begins with independently reproducing, but otherwise identical lineages differing in their initial value of p (Abrams 2006b).

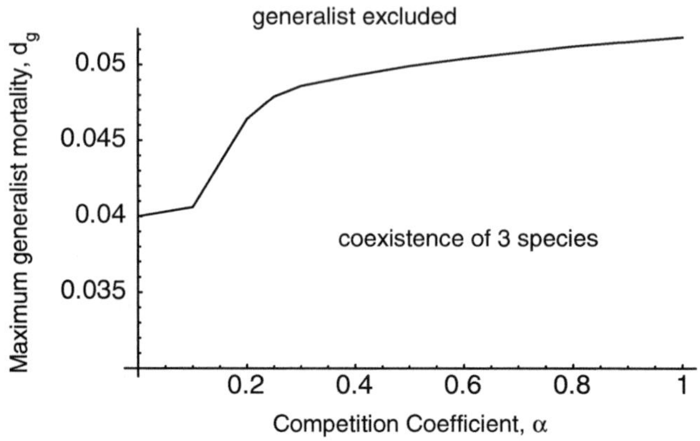

Figure 7.2. The maximum generalist death rates allowing coexistence of all three consumers in a system that is identical to that used in Figure 7.1, except for the amount of competition between resources (given on the x-axis), and the fact that the resource carrying capacities are increased by the factor $(1 + \alpha)$, to maintain a constant total resource density at equilibrium in the absence of consumers. The minimum generalist death rate allowing coexistence is the same as that of specialist 1, (i.e. $d = 0.03$) for all values of α, and is coincident with the x-axis.

As an illustration of the range of conditions allowing a generalist to persist with specialists, Figure 7.2 shows the maximum and minimum generalist death rates allowing three-species coexistence as a function of the amount of competition between the resources. (Except for resource competition, and a somewhat smaller resource immigration rate, this system is identical to that used in Figure 7.1, in which the generalist has an R^* equivalent to that of the specialists.) In the absence of the generalist, any competition between resources would change the indirect interaction between specialists from being nonexistent to being mutualistic. Resource competition also has a large affect on the temporal correlation of resource cycles, usually producing negatively correlated densities. Figure 7.2 shows that increased competition between resources has a positive effect on coexistence of three consumer species. The range of mortalities allowing the generalist to exist is almost twice as large as in a system with no competition for all α values greater than about 0.4. If the resources are themselves exploitative competitors with each other, these larger α values imply significant overlap in their use of different lower level resources (which are not represented explicitly here). It is worth noting that the results of this analysis and the subsequent ones are relatively insensitive to the magnitude of the resource immigration rate from outside the system (denoted I), provided that value is not large enough to prevent cycles. For example, increasing I by an order of magnitude from the level in Figures 7.1 and 7.2 (from 0.001 to 0.01) makes immigration equal to 1/25 the maximum possible resource growth rate; however, this reduces the range of death rates allowing coexistence by less than 20% for most of the competition coefficients shown in Figure 7.2.

7.3.2 Model 2: Adaptive change in the generalist's preferences

This section modifies Eqs. (7.1) by adding adaptive change in the generalist's choice parameter, p. Earlier work on similar systems had identified adaptive choice as being essential for obtaining coexistence (Wilson and Yoshimura 1994; Egas et al. 2004). This requirement appears to have arisen in those models due to the spatial separation of the two resources in two distinct habitats (Abrams 2006a). In any case, Abrams (2006a) showed that adding adaptive choice based on a trade-off in resource capture rates greatly expands the parameter range yielding coexistence for Eqs. (7.1), which assumes that there is a single homogeneous habitat. This section will briefly review those results, and extend them to systems in which there is competition between the two resources.

The basic assumption in this model is that the choice parameter p in Eqs. (7.1) changes at a rate proportional to its effect on the consumer's per capita growth rate. This can be described by:

$$\frac{dp}{dt} = \nu\left(\frac{dW}{dp}\right) + \frac{\varepsilon}{p^2} - \frac{\varepsilon}{(1-p)^2},$$

$$\text{where} \quad \frac{dW}{dp} = \frac{b_g n\left(p^{n-1}C_1R_1 - (1-p)^{n-1}C_2R_2\right)}{\left[1 + h\left(p^n C_1 R_1 + (1-p)^n C_2 R_2\right)\right]^2} \tag{7.2}$$

The use of this model in the context of behavioral choice is justified in greater detail in Abrams and Matsuda (2004) and Abrams (2006a,c). The constant ν translates the slope of the relationship between individual fitness and the choice trait (dW/dp) into a rate of change in p. This reflects the fact that behavioral change is observed to be more rapid when the fitness reward from a unit change in behavior is greater (e.g. Lester 1984). The final two terms of Eq. (7.2) represent the effects of two different processes that affect behavioral change when the behavior approaches a maximum or minimum value; the first is the fact that the magnitude of change in the direction of a more extreme value is limited. The second is that sampling behavior to assess the difference in fitness is likely to be biased towards the less extreme behaviour. Both of these have the effect of preventing change towards an extreme value, or equivalently, of biasing change towards less extreme behavior. The constant ε is assumed to be much smaller than 1, so these terms are only significant when p_i approaches its limiting values of 0 or 1. Although these final two terms in Eq. (7.2) are phenomenological, similar results have been obtained with several other similarly shaped functions of the mean trait, p. In the situations modeled here, the trade-off exponent, n, is assumed to be ≤ 1, so there is always a single fitness-maximizing value of p. Because of this, the consumer and resource dynamics predicted by combining Eqs. (7.1) and (7.2) are usually similar to those predicted in an alternative model in which p is assumed to instantaneously approximate its optimal value at each point in time. This is not surprising, because behavioral dynamics are assumed to be significantly faster than population dynamics. Given a large ν (appropriate for behavioral change), different behavioral types quickly converge to a common value of p, even if the currently optimal value of p undergoes temporal cycles with changing resource densities. Convergence means that it is reasonable to assume limited behavioral variation, and to characterize the p in the population by its mean value.

Abrams (2006a) shows that inclusion of behavioral switching by the generalist results in an approximate doubling of the range of generalist resource requirements (mortality rates) that allows three-species coexistence for a wide range of different combinations of the other parameters in Eqs. (7.1) and (7.2). The reason for this expansion is as follows. When the generalist mortality is close to the minimum that allows the specialists to exist, the generalist has a predominant influence on the dynamics of the resources. Rapid choice by the generalist tends to make the resource dynamics nearly synchronous when the resources differ only moderately in growth parameters. Under these circumstances, switches in preference have little effect on resource intake, so the minimum mortality is almost identical to the case where switching does not occur, and resources are synchronized simply by virtue of experiencing similar mortality effects from the generalist. When the resources differ greatly in their growth rate parameters, a switching generalist preys much less on the slower-growing resource than does a generalist with fixed preferences. As a result the adaptive species is less likely to cause exclusion of the specialist on the slower-growing resource, and the minimum mortality for coexistence given a rapidly switching adaptive generalist can then be smaller than for an inflexible generalist. On the other hand, the maximum mortality that the generalist can withstand is increased by behavioral switching. Switching clearly increases resource intake rate compared to fixed foraging, by allowing the generalist to increase its consumption of the currently most abundant resource. Greater intake in a given environment implies that a greater mortality rate can be tolerated.

Competition between resources appears to have a significant positive effect on the range of generalist mortality rates allowing coexistence with two specialists. In the example shown in Figure 7.3, the range of mortalities allowing coexistence expands from 0.03–0.0475 to 0.03–0.056 when the competition coefficients between the resource increase from 0 to 0.3. Further increase in competition increase this range slightly to 0.03–0.058 when α is increased from 0.3 to 1. This positive effect of between-resource competition on consumer coexistence is observed for the vast majority of parameter sets simulated. The shape of the maximum generalist mortality as a function of the competition coefficient is similar to the pattern shown in Figure 7.2 for the model with fixed consumer preferences. However, the maximum mortalities are higher in Figure 7.3. In both cases, the tendency of between-resource competition to produce offset peaks in the abundances of the two resources allows the generalist to persist when it has a greater mortality-rate disadvantage.

The preceding analysis and Abrams (2006a) both assume that the generalist has the same functional response and conversion efficiency parameters as does the specialist, so that if $p = 0$, the generalist is identical to specialist 2, and if $p = 1$, it is identical to specialist 1. However, this assumption is not the most conducive to coexistence with the specialists. In general, the more nearly linear the functional response

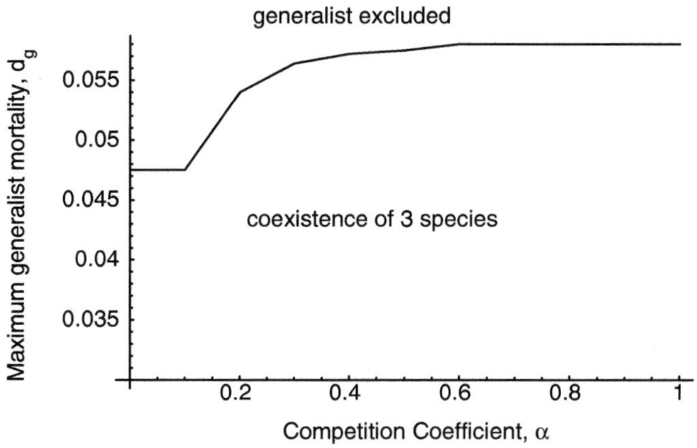

Figure 7.3. The range of generalist death rates allowing coexistence of all three consumers in a system with adaptive switching of consumption rates according to Eq. (7.2). The range of generalist death rates consistent with coexistence is shown for systems that vary in the amount of competition between resources. As in Figure 7.2, the baseline resource carrying capacities are increased by the factor $(1+\alpha)$, to maintain a constant total resource density at equilibrium in the absence of consumers. The trait dynamics parameters are: $\nu = 20$ and $\varepsilon = 0.0001$. The other parameters are identical to those used in Figure 7.2. As in Figure 7.2, the minimum generalist death rate allowing coexistence is the same as that of specialist 1, 0.03, for all α, and is coincident with the x-axis.

of the generalist, the larger the range of parameters (or R^* values) allowing it to coexist with the specialists. Having a more linear functional response enables the generalist to increase more rapidly during the periods when it has a greater resource intake than either specialist. Having a linear functional response would enable the generalist to make use of the same mechanism that allows coexistence of two species on a single resource in the Armstrong–McGehee mechanism (Armstrong and McGehee 1980; Abrams and Holt 2002). In some systems with between-resource competition, having linear functional responses allows an adaptively switching generalist to coexist with the specialists for most of the range of mortality rates that would allow it to exist in the absence of any competition from specialists (unpublished results).

The preceding discussion has dealt with a family of models under which the competition occurs in a single habitat and the environmental variation is endogenously generated consumer–resource cycles. In a similar analysis with noncompeting resources, Abrams (2006a) examined the impact on coexistence of cycles that were caused by temporal variation in resource growth parameters. Coexistence of two specialists and a generalist was shown to be possible in this case, but parameter ranges allowing this outcome were relatively restrictive unless the consumers had adaptive switching. With switching, coexistence occurred over a broad range of parameters when there was periodic variation in resource growth, and unpublished results suggest that the same is true of stochastic variation that is characterized by similar levels of autocorrelation. Preliminary results suggest that competition between the resources does not greatly alter these conclusions, but results will not be detailed here. Although the preceding analysis has assumed that there are only two resources, the impact of generalists on cycle synchrony and the effect of synchrony on coexistence apply equally well to systems in which each specialist consumes several resources, some or all of which are shared with the generalist. In these cases, the set of resources consumed by a specialist will cycle in synchrony with each other, and will behave approximately as a single resource.

7.4 HOW DO COEXISTING SPECIALISTS AND GENERALISTS IN VARIABLE ENVIRONMENTS RESPOND TO ALTERED MORTALITY OF ONE OF THESE SPECIES?

If the mechanism of coexistence described above operates in natural communities, then it is important to determine how such coexisting species would respond to environmental changes, such as increased mortality of one of them. Although this is a relatively simple food web, there are a variety of different indirect effects between species that make it difficult to predict the responses to mortalities or other environmentally influenced parameters without a mathematical analysis. Based on the results for models having more resources, linear functional responses, and stable equilibria, one might expect that, without resource competition, the generalist should have a negative effect on each specialist, so harvesting the generalist would always increase both specialists. Harvesting one specialist should increase the generalist, and therefore have a negative effect on the other specialist. When the resources compete, the two specialists have a mutualistic interaction in stable systems (Vandermeer 1980), and harvesting one should harm both. Logic from stable systems cannot be extended in any obvious manner to explain indirect effects when there is supersaturated coexistence and cycling populations. Moreover, Abrams et al. (2003) showed that a variety of counterintuitive changes in densities in response to mortality occurred in the "Armstrong–McGehee" system in which two consumers with different nonlinearities in their functional responses competed for a single resource. In that system both species could increase in average population size in response to mortality applied to one of them. It was also quite common for mean population sizes to exhibit discontinuous jumps with changes in the value of a parameter like mortality rate. Similar phenomena occur in the models described in the previous section.

The results below represent a very partial exploration of the types of responses to mortality that occur in the two classes of specialist–generalist communities described above. The facts that the models have a large number of parameters and that all results must be obtained by numerical integration mean that a complete catalogue of responses is not feasible. This section is primarily meant as a warning for those who might otherwise apply the logic appropriate for equilibrium systems to predict the impact of mortality rates in these cycling systems. The analysis is directed at the following questions: (1) How does the mortality rate of the generalist affect its own abundance and the abundances of the specialists with which it competes? (2) How does adaptive resource choice by the generalist consumer or competition between the resources alter the answer to question (1)? I also examine these same two questions for the case of mortality applied to one of the specialists. The answers are partial in that it was not possible to run simulations for the full range of ecologically possible parameter space. Nevertheless, the examples that are analyzed reveal a rather different set of responses than one might expect based on the logic of stable systems.

7.4.1 Models with inflexible foraging by the generalist

Figure 7.4 shows some examples of the changes in consumer densities as a function of generalist mortality in a system that has a generalist with inflexible resource choice, and can potentially support three consumer species. The two specialists have the same R^*, but differ by 20% in both their conversion efficiencies and mortality rates. This difference in vital rates prevents complete symmetry of the system, and gives the two specialist–resource systems different periods of population fluctuations in the absence of any coupling between them (such as competition between resources). In the absence of competition between resources (Figure 7.4A), there is a rapid drop in the mean density of the generalist by close to a factor of 6 as its mortality increases enough to allow the two specialists to enter the system. The generalist density when its mortality is just to the left of the y-axis in Figure 7.4A (i.e. $d_g = 0.0299$) is 0.597, and the two specialist densities drop to zero. Once mortality of the generalist is within the range allowing coexistence, increased mortality usually reduces the generalist density and increases the specialist populations, but these changes are quite small in magnitude over most of the range of mortality rates allowing coexistence. The qualifier "usually" is present in the preceding sentence because generalist's density actually increases when its mortality is slightly less than the value that causes generalist extinction. The generalist then drops to zero density over a very small interval of mortality rates. The two specialists are relatively insensitive to generalist mortality except for the threshold values at which the generalist either no longer causes specialist exclusion or is itself excluded.

Competition between the resources increases the density of the generalist in systems where the three consumers coexist. As was shown earlier, increasing the strength of competition between the resources increases the range of generalist mortality rates that allow coexistence. Figure 7.4 illustrates the mortality vs. abundance relationships that occur with increasing strength of competition between the resources. An intermediate strength of competition, such as that shown in Figure 7.4C, results in very irregular patterns of changes in consumer densities with increasing generalist mortality. Some of these transitions are due to qualitative shifts in the system dynamics (e.g. between cycles and chaos, or between simple and complex cycles) as the mortality rate parameter is changed. For example, the lower range of generalist mortalities illustrated in Figure 7.4C involve very complex, long-period cycles or chaos, while simple cycles with relatively short periods

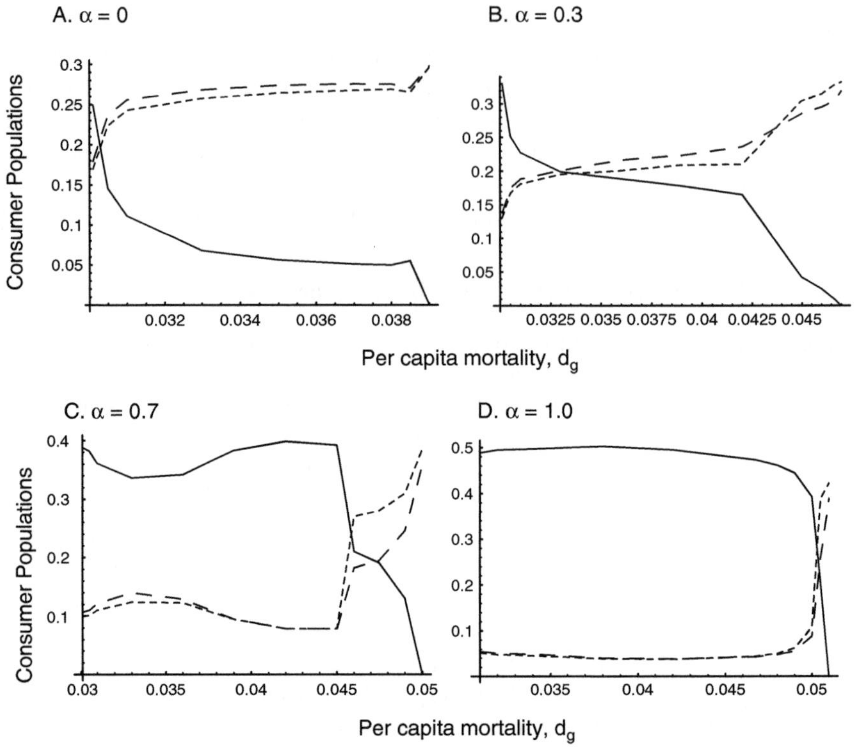

Figure 7.4. Four panels showing how generalist mortality affects the abundance of a generalist consumer having fixed preferences and two specialist consumers for different levels of competition between the resources. The value of the generalist's choice parameter p is fixed at 0.5. Panels a through d show densities for systems having increasing competition between resources. The other parameter values for all systems are: $r_1 = r_2 = 1$; $K_1 = K_2 = 1 + \alpha$; $I_1 = I_2 = 0.001$; $b_1 = b_g = 0.1$; $b_2 = 0.12$; $d_1 = 0.03$; $d_2 = 0.036$; $C_1 = C_2 = 5$; $h = 1$. In this and all subsequent figures, the generalist is denoted by the solid line, the specialist on resource 1 by the short-dashed line, and the specialist on resource 2 by the long-dashed lines.

predominate at higher generalist mortalities. Figure 7.4 also shows that increasing competition between resources (without changing the total equilibrium resource abundance) increases the abundance of the generalist and decreases the abundances of specialists. This appears to be primarily a consequence of the desynchronizing effect of the competition, as explained in Vandermeer (2004). Decreased temporal synchrony in the fluctuations of the two resources favors generalists over specialists. It is interesting to note that, in the absence of the generalist, increasing competition between resources increases both specialist populations (as noted for stable systems by Vandermeer 1980).

The above simulations assumed that the generalist was equally adept at catching both resources. In other simulations in which $p \neq 0.5$, the results were changed remarkably little. If $p = 0.25$ or $p = 0.75$, the range of mortalities allowing coexistence was reduced somewhat, but the general pattern of responses of densities to generalist mortality shown in Figure 7.4 was changed only slightly. Other simulations in which the two resources differed in the intrinsic growth rates but specialist consumers had equal demographic parameters also yielded results qualitatively similar to those shown in Figure 7.4.

Figure 7.5 shows the corresponding analysis of how the mortality rate of one of the specialists affects the mean population densities of both the generalist and the other specialist. Parameter values are identical to those assumed in Figure 7.4, but the mortality rate of specialist 1 rather than the generalist is varied, and a different set of resource competition coefficients is shown. This figure exhibits the same general features as does Figure 7.4. For a broad range of mortalities allowing coexistence, there is very little impact of mortality applied to one species, either on its own average density or on the densities of the species with which it interacts. However, when the mortality of one specialist becomes sufficiently high, that specialist declines to an equilibrium abundance of zero over a very narrow range of mortality rates. This is accompanied by a huge increase in the generalist, and a large decrease (but not exclusion) of the other specialist. If the direction of change in mortality is reversed there will then be a large decrease in generalist abundance and increase in the resident specialist abundance when the mortality of the formerly excluded specialist is reduced enough to allow it to reinvade. Figure 7.5 also shows the same large impact of between-resource competition on the relative densities of the generalist and specialists as did Figure 7.4, with the generalist dominating at high resource competition, and specialists dominating at low levels of resource competition.

7.4.2 Models with behavioral switching between resources by the generalist

Analyses similar to the preceding ones were carried out to determine the impact of mortality of either the generalist or one specialist on all population densities in comparable systems where the generalist exhibited adaptive switching. In the numerical results reported here, I assume a linear trade-off between the two attack rates; $n = 1$. The asymmetry in this example was due to unequal intrinsic growth rates of the two resources. The consumer population sizes vs. generalist mortality for three levels of competition between resources are shown in Figure 7.6. As in the case of inflexible consumer behaviour, competition between resources determined whether specialists or generalists were more abundant, with high competition favoring the generalist. However, the generalist never attains the large population densities at high levels of resource competition that occur in the absence of behavioral switching. Behavioral choice led to many cases of complex dynamics, and alternative attractors existed for ranges of several parameters in most of the systems explored. (However, in most cases, these either existed over relatively small ranges of mortality rates or did not have a very large effect on average densities.) Although the parameters are not identical between Figures 7.4 and 7.6, the general patterns exhibited by Figure 7.6 are

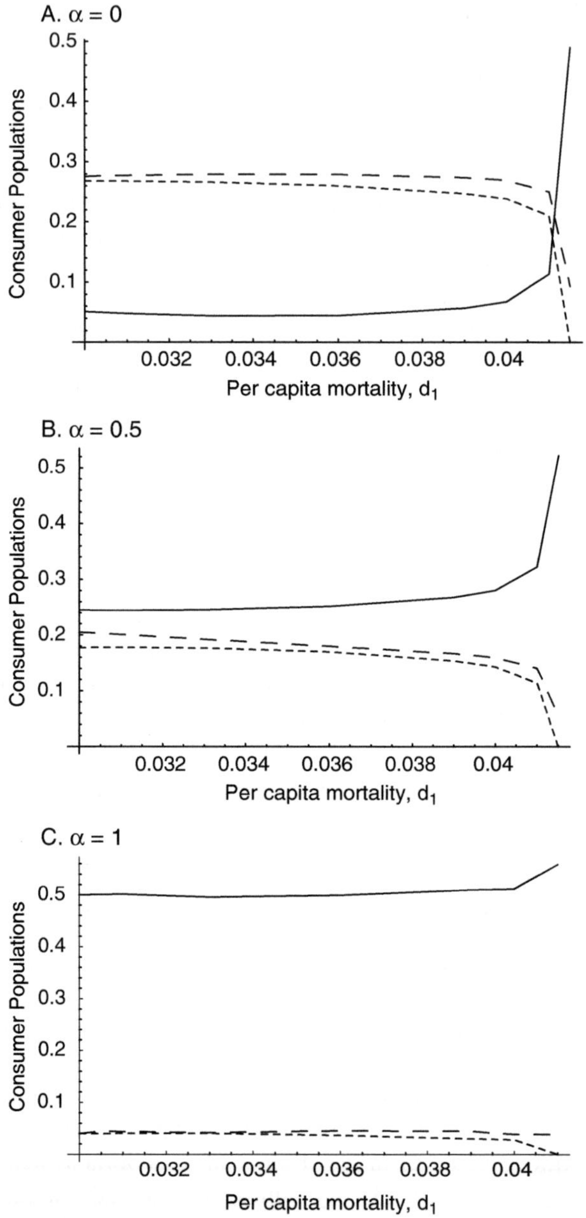

Figure 7.5. Three panels showing how mortality on one specialist affects the abundances of the generalist and the other specialist in a system identical to that considered in Figure 7.4, with the generalist mortality set at $d_g = 0.03$. Line styles denoting the three consumers are as in Figure 7.4.

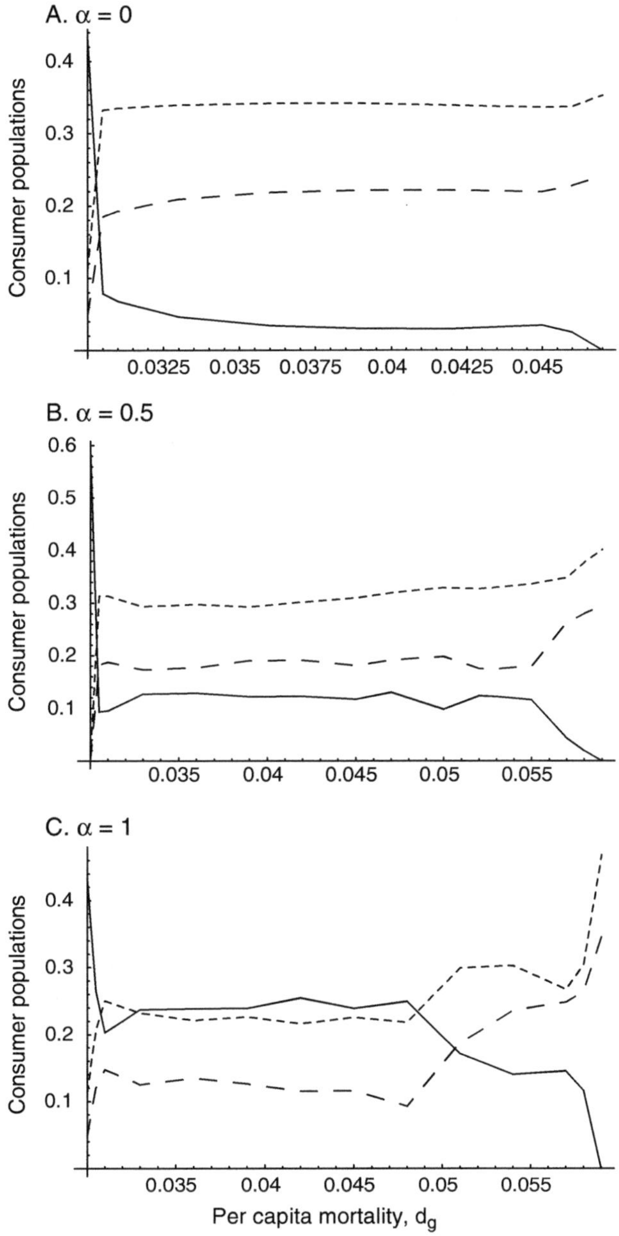

Figure 7.6. Three panels showing how mortality of a switching generalist affects the abundances of all of the consumer species. The dynamics are based on Eqs. (7.1) and (7.2) with parameter values: $r_1 = 1.2$; $r_2 = 0.8$; $K_1 = K_2 = 1 + \alpha$; $I_1 = I_2 = 0.001$; $b_1 = b_g = b_2 = 0.1$; $d_1 = d2 = 0.03$; $C_1 = C_2 = 5$; $h = 1$; $v = 20$; $n = 1$; $\varepsilon = 0.0001$. Line styles denoting the three consumers are as in Figure 7.4.

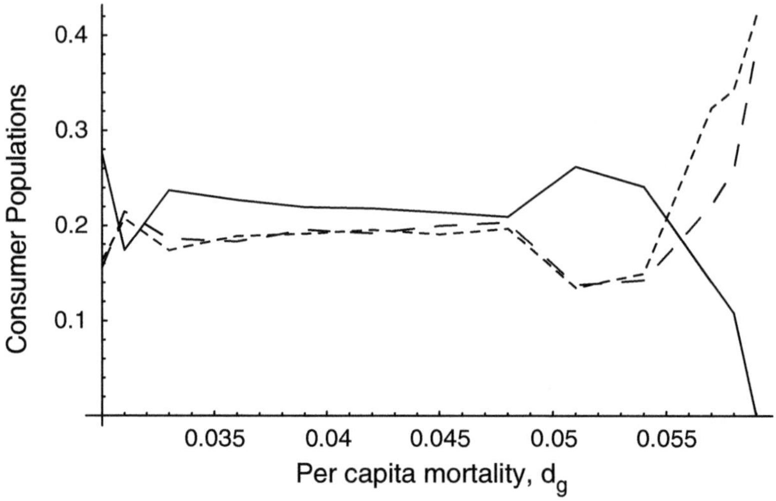

Figure 7.7. An illustration of the effect of mortality applied to a switching generalist in a system where the two specialists differ only in their demographic response rates. The competition between resources is complete ($\alpha = 1$), and the other parameters are: $r_1 = 1$; $r_2 = 1$; $K_1 = K_2 = 2$; $I_1 = I_2 = 0.001$; $b_1 = 0.1$; $b_2 = 0.12$; $d_1 = 0.03$; $d_2 = 0.036$; $C_1 = C_2 = 5$; $h = 1$. Line styles denoting the three consumers are as in Figure 7.4.

changed very little when all parameters other than switching are identical to those in Figure 7.4. This generalization is reflected in Figure 7.7 which assumes complete resource competition ($\alpha = 1$) but uses the other parameters assumed in Figure 7.4D. Comparing Figures 7.4D and 7.7 shows that the ability to switch leads to much lower generalist densities, higher specialist densities, and a more complex trajectory of mean population sizes vs. generalist mortality rate.

When there is high resource competition, the lower abundance of the generalist in models with switching than in models with fixed behaviour (Figure 7.7 vs. 7.4D) appears to be due to the increased overexploitation of resources allowed by switching. However, this overexploitation does not translate into a larger effect of generalists on specialists in the models with switching than those without. The abundances of specialists are considerably greater in the models with complete competition between resources when switching is present than when it is not. This appears to be due to the lag in switching by the generalist, which allows more rapid growth of the specialist when it is recovering from low population size. The increased dynamic complexity caused by switching leads to repeated shifts between qualitatively different types of dynamics (simple cycles, complex cycles, chaos) as the mortality parameter increases. This makes the shape of the relationship between mortality and population size considerably more irregular than in the case of consumers having inflexible behaviour. Figure 7.8 provides an example of the impact of

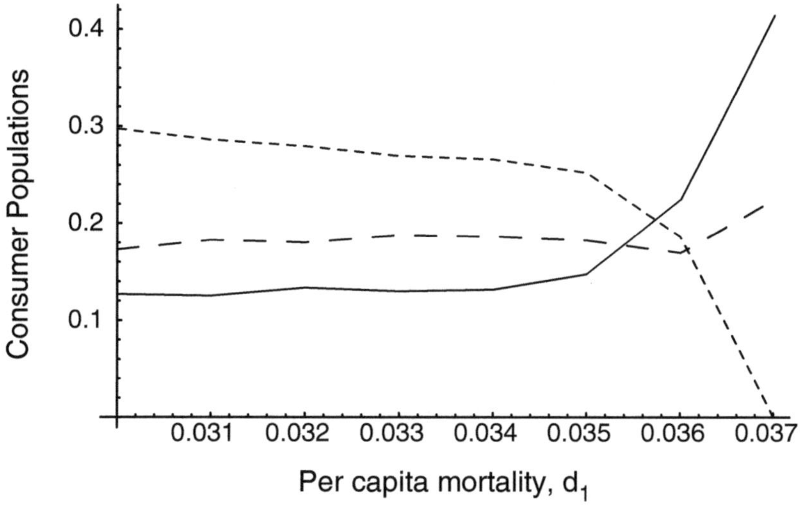

Figure 7.8. The effect of the mortality rate of specialist 1 on the abundances of all of the consumer species in the system illustrated in Figure 7.6B ($\alpha = 0.5$) for a generalist mortality of $d_g = 0.036$. Line styles denoting the three consumers are as in Figure 7.4.

one of the specialist's mortality rates on mean population densities in a system with moderate competition between resources. The model is equivalent to Figure 7.6B in all parameters other than consumer mortality rates. As is true of increasing the mortality of the generalist (Figure 7.6B), increasing specialist mortality (Figure 7.8) has relatively little impact on the mean population sizes of any of the consumers over a broad range of mortality rates. However, as in the comparison involving generalist mortality (Figures 7.4 vs. 7.6), switching generalist consumers (Figure 7.8) are more responsive to specialist mortality than are nonswitching consumers (Figure 7.5B).

There is clearly a great deal more that needs to be learned about competition in variable systems. This generalization also applies to systems in which the number of consumers does not exceed the number of resources. For example, there has not been any analysis of the impact of one specialist's mortality on the two specialist densities in systems that lack the generalist. This leaves open the possibility that supersaturation is not an essential aspect of the results presented in this section. However, limited numerical work supports the conjecture that this is not the case. Figure 7.9 is an example of the responses of the two specialists to mortality of specialist 1 in the system investigated in Figure 7.7D after removing the generalist. As one would expect based on the mutualistic indirect effect between these two species, increasing the mortality of one specialist generally decreases both. However, it is clear that this generalization does not apply over the entire range of potential mortality rates; competition between consumers reemerges at high mortality rates.

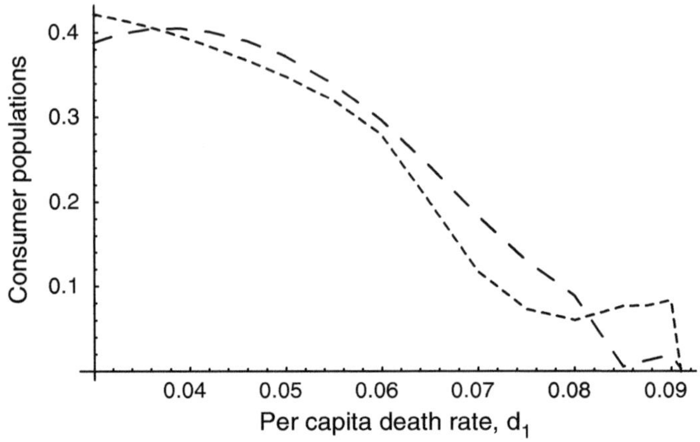

Figure 7.9. An illustration of the effect of mortality on one specialist on both specialist densities in a system in which there is no generalist. The competition between resources is complete ($\alpha = 1$), and the other parameters are: $r_1 = 1$; $r_2 = 1$; $K_1 = K_2 = 2$; $I_1 = I_2 = 0.001$; $b_1 = 0.1$; $b_2 = 0.12$; $d_1 = 0.03$; $d_2 = 0.036$; $C_1 = C_2 = 5$; $h = 1$. Line styles denoting the two specialist consumers are as in Figure 7.4.

7.5 DISCUSSION

The most important messages from the work reported above are: (1) coexistence of one generalist and two specialists on two cycling resources could occur in a wide range of systems, particularly in cases where the generalist (or all of the species) can exhibit adaptive choice of resources; (2) coexistence is usually made more likely by the presence of competition between the two resources that are the subject of competition; and (3) the densities of the competing consumers can respond in very nonlinear and often counterintuitive ways to mortality imposed on one of the species. One of the disturbing possibilities exhibited by most of the models considered here is that mean consumer population sizes are very insensitive to changing mortality rates over a broad range of mortalities, but can abruptly go extinct when mortality surpasses a threshold value. This pattern, or at least very rapid declines over a narrow range of mortalities are characteristics of all of the supersaturated systems examined here. This possibility makes it important to identify situations in which the mechanism described here might be a major factor contributing to coexistence. While there are certainly a number of candidate systems in which both consumer–resource cycles and competition between consumers occur (e.g. the predators on voles (Hanski et al. 2001) or those on snowshoe hare (Krebs et al. 2001)), we do not have sufficient information to judge whether these cycles contribute to the existence or abundance of generalists. Systems with microbial competitors may be among the most tractable for addressing issues of specialist–generalist coexistence (Kassen 2002).

The results reported here might seem to support the notion that coexistence of species competing for substitutable resources that vary because of endogenous cycles is usually likely to involve only one more consumer species than the number of resources. However, the mechanism explored here – basically the generalist consuming negative covariance in resource densities – is only one of several possible mechanisms that might operate concurrently. If each of the specialist–resource systems were replaced by two specialists that differ greatly in the nonlinearity of their functional responses, as proposed by Armstrong and McGehee (1980), it seems likely that the number of robustly coexisting species could easily be expanded to 5. And it is certainly possible that models with larger numbers of resources could allow many more consumer species than resources to coexist.

The results reported here add to a growing list of cases in which interactions between species in variable environments are fundamentally different from those in constant environments. Most research with this general message has studied resource competition, but a growing body of literature shows that a similar general message applies to the case of apparent competition between prey species (Abrams et al. 1998; Abrams 1999; Brassil 2006), and to three-species food chains (Abrams and Roth 1994).

There is still much that needs to be studied in the area of competition between species in variable environments. This chapter has concentrated on variation that arises from consumer–resource cycles. It is known that mechanisms similar to those discussed here allow coexistence of specialist and generalist species consuming resources that fluctuate in density because of temporal variation in resource growth parameters (Abrams 2006a). Preliminary work suggests that competitive interactions in these cases are also quite different from those expected when coexistence occurs purely via resource partitioning in either constant or cycling systems.

7.6 ACKNOWLEDGMENTS

My first encounter with Peter Yodzis' work was his 1977 American Naturalist article on the limiting similarity of competing species. It is therefore a pleasure to be able to contribute a work on the factors limiting coexistence of competitors to this volume. I thank the Natural Sciences and Engineering Research Council of Canada for financial support, and thank Claus Rueffler and an anonymous reviewer for helpful comments on an earlier draft.

7.7 LITERATURE CITED

Abrams, P.A. 1999. Is predator mediated coexistence possible in unstable systems? *Ecology* 80: 608–621.

Abrams, P.A. 2004. When does periodic variation in resource growth allow robust coexistence of competing consumer species? *Ecology* 85: 372–382.

Abrams, P.A. 2006a. The prerequisites for and likelihood of generalist-specialist coexistence. *American Naturlalist* 167: 329–342.

Abrams, P.A. 2006b. Evolution of resource-exploitation traits in a generalist consumer: the evolution and coexistence of generalists and specialists. *Evolution* 60: 427–439.

Abrams, P.A. 2006c. The effects of switching behavior on the evolutionary diversification of generalist consumers. *American Naturalist* 168: 645–659.

Abrams, P.A., C.E. Brassil, and R.D. Holt. 2003. Dynamics and responses to mortality rates of competing predators undergoing predator-prey cycles. *Theoretical Population Biology* 64: 163–176.

Abrams, P.A. and R.D. Holt. 2002. The impact of consumer-resource cycles on the coexistence of competing consumers. *Theoretical Population Biology* 62: 281–295.

Abrams, P.A., R.D. Holt, and J.D. Roth. 1998. Shared predation when populations cycle. *Ecology* 79: 201–212.

Abrams, P.A. and H. Matsuda. 2004. Consequences of behavioral dynamics for the population dynamics of predator-prey systems with switching. *Population Ecology* 46: 13–25.

Abrams, P.A. and Roth, J. 1994. The effects of enrichment on three-species food chains with nonlinear functional responses. *Ecology* 75: 1118–1130.

Armstrong, R.A. and R. McGehee. 1976a. Coexistence of two competitors on one resource. *Journal of Theoretical Biology* 56: 499–502.

Armstrong, R.A. and R. McGehee. 1980. Competitive exclusion. *American Naturalist* 110: 151–170.

Brassil, C.E. 2006. Can environmental variation generate positive indirect effects in a model of shared predation? *American Naturalist* 167: 43–54.

Chesson, P. 1994. Multispecies competition in variable environments. *Theoretical Population Biology* 45: 227–276.

Chesson, P. 2000. Mechanisms of maintenance of species diversity. *Annual Review of Ecology and Systematics* 31: 343–366.

Coste, J., J. Peyraud, P. Coulet, and A. Chenciner. 1978. About the theory of competing species. *Theoretical Population Biology* 14: 165–184.

Ebenhöh, W. 1988. Coexistence of an unlimited number of algal species in a model system. *Theoretical Population Biology* 34: 130–144.

Egas, M., U. Dieckmann, and M.W. Sabelis. 2004. Evolution restricts the coexistence of specialists and generalists: the role of trade-off structure. *American Naturalist* 163: 518–531.

Futuyma, D.J. and G. Moreno. 1988. The evolution of ecological specialization. *Annual Review of Ecology and Systematics* 19: 207–234.

Hanski, I., H. Hettonen, E. Korpimäki, L. Oksanen, and P. Turchin. 2001. Small rodent dynamics and predation. *Ecology* 82: 1505–1520.

Holling, C.S. 1959. The components of predation as revealed by a study of small mammal predation of the European pine sawfly. *Canadian Entomologist* 91: 293–320.

Huisman, J. and F.J. Weissing. 1999. Biodiversity of plankton by species oscillations and chaos. *Nature* 402: 407–410

Huisman, J. and F.J. Weissing. 2001. Biological conditions for oscillations and chaos generated by multispecies competition. *Ecology* 82: 2682–2695.

Huisman, J. and F.J. Weissing. 2002. Oscillations and chaos generated by competition for interactively essential resources. *Ecological Research* 17: 175–181.

Hsu, S.B., S.P. Hubbell, and P. Waltman. 1978. A contribution to the theory of competing predators. *Ecological Monographs* 48: 337–349.

Kassen, R. 2002. The experimental evolution of specialists, generalists, and the maintenance of diversity. *Journal of Evolutionary Biology* 15: 173–190.

Kendall, B.E., J. Prendergast, Bjornstad. 1998. The macroecology of population dynamics: taxonomic and biogeographic patterns in population cycles. *Ecology Letters* 1: 160–164.

Kisdi, E. 2002. Dispersal: risk spreading versus local adaptation. *American Naturalist* 159: 579–596.

Koch, A.L. 1974. Competitive coexistence of two predators utilizing the same prey under constant environmental conditions. *Journal of Theoretical Biology* 44: 387–395.

Krebs, C.J., S. Boutin, and R. Boonstra. 2001. *Ecosystem dynamics in the boreal forest*. Oxford University Press, Oxford.

Lawton, J.H., T.M. Lewinsohn, and S.G. Compton. 1993. Patterns of diversity for the insect herbivores on bracken. pp. 178–184, in R.E. Ricklefs and D. Schluter, *Species diversity in ecological communities: historical and geographical perspectives*. University of Chicago Press, Chicago.

Leon, J.A. and D. Tumpson. 1975. Competition between two species for two complementary or substitutable resources. *Journal of Theoretical Biology* 50: 185–201.

Lester, N.P. 1984. The feed-feed decision: how goldfish solve the patch depletion problem. *Behaviour* 89: 175–199.

Levins, R. 1979. Coexistence in a variable environment. *American Naturalist* 114: 765–783.

Litchman E. and C.A. Klausmeier. 2001. Competition of phytoplankton under fluctuating light. *American Naturalist* 157: 170–187.

May, R.M. and W.J. Leonard. 1975. Nonlinear aspects of competition between three species. S.I.A.M. *Journal of Applied Mathematics* 29: 243–253.

Murdoch, W.W. 1969. Switching in generalist predators: experiments on prey specificity and stability of prey populations. *Ecological Monographs* 39: 335–354.

Nakajima, M., H. Matsuda, and M. Hori. 2004. Persistence and fluctuation of lateral dimorphism in fishes. *American Naturalist* 163: 692–698.

Parvinen, K. and M. Egas. 2004. Dispersal and the evolution of specialisation in a two-habitat type metapopulation. *Theoretical Population Biology* 66: 233–248.

Ruel, J.J. and M.P. Ayres. 1999. Jensen's inequality predicts effects of environmental variation. *Trends in Ecology & Evolution* 14: 361–366.

Schippers, P., A.M. Verschoor, M. Vos, and W.M. Mooij. 2001. Does "supersaturated coexistence" resolve the "paradox of the plankton"? *Ecology Letters* 4: 404–407.

Stewart, F.M. and B.R. Levin. 1973. Partitioning of resources and the outcome of interspecific competition: a model and some general considerations. *American Naturalist* 107: 171–198.

Vandermeer, J.H. 1980. Indirect mutualism: variations on a theme by Stephen Levine. *American Naturalist* 116: 440–448.

Vandermeer, J. 2004. Coupled oscillations in food webs: balancing competition and mutualism in simple ecological models. *American Naturalist* 163: 857–867.

Wilson, D.S. and J. Yoshimura. 1994. On the coexistence of generalists and specialists. *American Naturalist* 144: 692–707.

Yodzis, P. 1977. Harvesting and limiting similarity. *American Naturalist* 111: 833–843.

CHAPTER 8

ENVIRONMENTAL VARIABILITY MODULATES THE INSURANCE EFFECTS OF DIVERSITY IN NON-EQUILIBRIUM COMMUNITIES

ANDREW GONZALEZ*

Department of Biology, McGill University, 1205 ave Docteur Penfield, Montreal, QC H3A 1B1, Canada

**Corresponding author A. Gonzalez Phone: 514-398-6444, Fax: 514-398-5069, E-mail: andrew.gonzalez@mcgill.ca*

OSCAR DE FEO

Department of Microelectronic Engineering, University College Cork, Butler Building, North Mall, Cork, Ireland

8.1 ABSTRACT

Previous theoretical analyses of the relation between diversity and community stability have ignored or greatly simplified the environment. One ubiquitous feature of stochastic environments is the autocorrelation, or long-term memory, that is apparent over multiple time scales. We analyze a model of nonequilibrium resource competition and show how environmental autocorrelation can strongly define community stability. Autocorrelated environmental variability can have destabilizing

David A. Vasseur and Kevin S. McCann, The Impact of Environmental Variability on Ecological Systems, 159–177.

effects 50 times greater than the stabilizing effects of increasing species richness; this result was exacerbated when we considered a case where increasing species richness had a destabilizing effect on community dynamics. In particular, the interaction between nonlinear resource competition and autocorrelation in the environment alters the magnitude and timing of population and community fluctuations. Increasing the autocorrelation of the environment caused a qualitative shift in community dynamics from a single long-term "steady state", in which population fluctuations were small and biomass was relatively evenly distributed across all species, to metastable dynamics characterized by periods of relative stability, punctuated by abrupt shifts in species dominance and a highly uneven biomass distribution. Consistent with recent metacommunity theory the addition of immigration stabilized the community. We suggest that increasing autocorrelation, due to climate change, may have hitherto unforeseen destabilizing effects on community dynamics that may compound the effects of ongoing species loss.

Keywords: environmental fluctuations, autocorrelation, diversity, stability, community evenness, resource competition, dispersal.

8.2 INTRODUCTION

Ecological communities are perturbed by environmental fluctuations over a great range of temporal scales (Davis 1986; Levin 1992; Roy et al. 1996; Bennett 1997; Shugart 1998; Dynesius and Jansson 2000). Differential species responses to short and long-term environmental variation ensure that communities are in a constant floral and faunal disequilibrium. The differences between species in their capacity to respond to environmental variation causes some species to track change closely, whilst others respond much more slowly, resulting in time-lagged responses to environmental change occurring over much longer timescales (Davis 1986; Bennett 1997).

The shifts in distribution and abundance that accompany species responses to environmental change can have profound affects upon community structure and functioning (Steele and Henderson 1984; Braswell et al. 1997; McGowan et al. 1998; Hulme et al. 1999; Dynesius and Jansson 2000; Knapp et al. 2002; Stenseth et al. 2002). Recent ecological theory suggests that short-term variation in species abundance may affect the stability of community processes. Specifically, aggregate community processes, such as biomass production, will be stabilized by differences in the magnitude and timing of species responses to the environment, the asynchronous fluctuations at the population level result in a buffering effect at the community level (Doak et al. 1998). Furthermore, increasing species diversity is thought to enhance this buffering effect by increasing the range of species responses to environmental fluctuations (McNaughton 1977; Doak et al. 1998; Hughes and Roughgarden 1998; Ives et al. 1999; Yachi and Loreau 1999; Chesson et al. 2001; Norberg et al. 2001; Ives and Hughes 2002).

Ives and Hughes (2002) define two conditions necessary for a stabilizing effect of increasing species richness. Firstly, the net effect of competition on the variability in population growth rates of component species should be independent of species number (i.e. variability in per capita growth rates should remain the same irrespective of the number of species in the community) and the strength of interspecific competition. Secondly, all species within the community should have approximately equal densities. Although these assumptions are not excessively unrealistic, conditions under which they break down are known. For example, recent experimental tests of this theory (Petchey et al. 2002; Gonzalez and Descamps-Julien 2004) suggest that the variability of population growth rates is not independent of species richness. Furthermore, community abundance distributions are not uniform but rather are typified by the relatively long-term dominance of one or a few species and rare species are maintained by immigration, a characteristic of natural communities that would seem to be the norm rather than the exception (e.g. Rahel 1990; Venrick 1990; Bengtsson 1994).

Another aspect that may limit the generality of recent theoretical results is the observation that the conditions for stability have been based on linearization techniques. The assumptions of the linear approach are that the community must be close-to-equilibrium, that environmental fluctuations are of small amplitude, and that the system must not be close to a bifurcation point. Yet, much of the literature indicates that long-term community dynamics, whether aquatic or terrestrial, are typified by large amplitude fluctuations that are driven by environmental variation (e.g. Willis et al. 1995; Whitlock and Bartlein 1997; Francis et al. 1998; Allen et al. 1999). This apparent inconsistency between theoretical studies and observations suggests that studies of the stability of far-from-equilibrium communities would be worthwhile. We address this issue, and the others raised earlier, by examining whether previous conclusions regarding the stabilizing effect of species diversity are altered by considering large amplitude environmental fluctuations, and in particular, whether the scale of autocorrelation of these fluctuations can affect community stability.

The temporal scaling of environmental variability can be characterized by its frequency structure. For example, temperature fluctuations in marine and terrestrial environments tend to increase in power (variance) at low frequencies (Mandelbrot and Wallis 1969; Steele 1985; Pelletier 2002; Vasseur and Yodzis 2004), by analogy with light such spectra are referred to generically as "red" as opposed to "white" spectra, in which all frequencies have equal variance. The property of increasing power (variance) through time is responsible for the long-term autocorrelation structure, or "memory" typical of many environmental variables. Previous work has identified the temporal scaling of environmental fluctuations as a key factor that may affect many ecological processes. Theory and experiments have demonstrated that extinction probabilities (Ripa and Lunderg 1996; Petchey et al. 1997; Cuddington and Yodziz 1999; Halley and Kunin 1999; Wichmann et al. 2003) as well as various aspects of population (Steele and Henderson 1984; Heino 1998; Morales 1999; Petchey 2000; Gonzalez and Holt 2002; Inchausti and Halley 2002) and community dynamics (Caswell and Cohen 1995; Ripa et al. 1998; Ripa and Ives 2003; Ripa and Ives, this volume) are dependent upon the autocorrelation in the environment.

To examine whether community stability is affected by the temporal scaling of the environment we adopt a standard resource competition model where species coexistence is fluctuation-dependent (*sensu* Chesson 2000) and is made possible by a storage effect (Lehman and Tilman 2000; Chesson et al. 2001). The buffered population growth generated by this model is also the basis of the stabilizing (insurance) effect of diversity previously reported (Ives et al. 1999; Yachi and Loreau 1999; Chesson et al. 2001; Norberg et al. 2001; Ives and Hughes 2002), what we are concerned with here is establishing how fluctuations in some environmental variable, other than the resource, modulate this buffering effect. Recent work has shown that "opening" a community to immigration can qualitatively alter community dynamics and stability (Holt et al. 2003; Loreau et al. 2003) and so we also study the effect of environmental scaling in the presence of immigration. Consistent with earlier results we show that diversity has a stabilizing effect on community biomass but that the temporal scaling of the environment *is* an important determinant of long-term population and community stability. In particular, we show that reddened environmental fluctuations can generate strong asymmetries in community structure that result in low community stability despite the presence of compensatory fluctuations between species. In addition, to meet the challenge of recent experimental results that suggest a destabilizing role of diversity (Petchey et al. 2002; Gonzalez and Descamps-Julien 2004; Petchey this volume) we analyze a case where the breadth of species' environmental tolerance (realized niche) is dependent upon species diversity. This follows from the basic expectation that the realized niche of any organism will be smaller than its fundamental niche (environmental tolerance) due to biotic interactions such as competition (Austin et al. 1990). Consistent with experimental results we demonstrate that this assumption can generate a destabilizing effect of increasing species richness. However, over long timescales community stability is still primarily dependent upon the temporal scaling of the environment.

8.3 MODEL

We consider a standard resource competition model and, to provide a comparison with previous work, we follow the parameterization of Lehman and Tilman (2000). The model is written in the following form (see Appendix 8A):

$$\begin{cases} \dot{R} = 1 - R - \sum_{i=1}^{N} g_i(x)\frac{RB_i}{R+K_i} \\ \dot{B}_i = B_i\left[g_i(x)\frac{R}{R+K_i} - m_i\right] + I_i \quad i \in \{1, 2, \ldots, N\} \\ g_i = r_i \exp\left[\frac{1}{2}\left(\frac{x-\tau_i}{w_f/N^\beta}\right)^2\right] \end{cases} \tag{8.1}$$

where N is the number of species, R is the amount of resource, B_i is the biomass of species i, K_i is the half saturation constant of the resource limitation function of species i, m_i is the specific mortality rate of species i, I_i is an exogenous constant immigration flow, $x \in [-1, 1]$ is the environmental variable and τ_i is the value of x for which species i has its optimal productivity rate, which falls off as a Gaussian curve away from this value, β is a scaling constant that determines how the width of species environmental tolerance w_f is altered by increasing species richness. Written in this form time is now measured in resource time constant units, providing a relative measure of all the frequencies of the biomass and environmental fluctuations with respect to the resource. For every simulation run the m_i, K_i, w_f and r_i were identical and equal to 0.1, 0.1, 0.2 and 1.0, respectively, and consistent with previous analyses (Lehman and Tilman 2000). We present the four cases of $\beta = 0$ and $\beta = 0.2$, together with $I_i = 0$ and $I_i = 9.90\text{E} - 4$. A $\beta = 0$ assumes that the width of the environmental tolerance (fundamental niche) of each species is unaltered by increasing diversity, whilst a $\beta = 0.2$ assumes a weak monotonic and symmetrical decline in the width of the environmental tolerance with increasing species diversity (i.e. a decline in the realized niche ($w = \frac{w_f}{N^\beta}$) of each species with species addition). $I_i = 0$ assumes that there is no exogenous immigration and is consistent with previous analyses (Lehman and Tilman 2000) whilst $I_i = 9.90\text{E} - 4$ assumes an exogenous immigration flow of 0.01% of the maximal single species density achievable (see Appendix 8A). The additions of β and I_i to this model are new and provide a simple, and plausible, means of studying cases where the addition of species may have negative effect upon community stability (e.g. Petchey et al. 2002; Gonzalez and Descamps-Julien 2004), and the effects of a continuous flow of immigrants (a "mass effect") from outside the local community (e.g. Schmida and Wilson 1985). Because of the absence of demographic stochasticity, population biomass can reach very low levels without suffering extinction; the effect of this potentially unrealistic assumption is assessed by studying the effects of immigration that prevents populations from attaining very low levels of biomass.

We vary the environmental variable x as a $1/f$ process (Halley 1996), i.e. a process where the spectrum relating the frequency, f, and power $S(f)$ is of the following form:

$$S(f) \propto \frac{1}{f^\gamma} \tag{8.2}$$

Typically, for natural environmental variables $\gamma \in [0.5, 1.5]$ and they are therefore reddened (e.g. Pelletier 2002). Thus the power (amplitude) is directly proportional to the power γ of the reciprocal of frequency and low frequency dominance engenders the property of increasing variance and a long-term correlation structure.

A $1/f$ time series can be approximated by the following function:

$$x(t) = \sum_{f=1}^{n/2} \frac{1}{f^{\frac{\gamma}{2}}} \sin\left(\frac{2\pi f t}{n} + \theta_f\right) \tag{8.3}$$

where n is the length of the series, f the frequency, t is the time, γ determines the relation between power and frequency and θ_f is a uniform deviate in $[0, 2\pi)$ which adds random phase to each sine wave. We simulated environmental fluctuations with $n = 512$ and $\gamma \in [0, 2]$, replicate series with the same mean and variance were constructed, and rescaled in the interval $[-1, 1]$, for each simulation run using a method of spectral mimicry (Cohen et al. 1999).

We varied species richness from 2 to 24 and examined population and community stability, in response to changing γ, by extensive numerical simulation. The model was numerically integrated using the Heun algorithm (Quarteroni 1999) with a fixed time step ($dt = 0.02$) for 60,000 units of time, where the first 10,000 were dropped and the statistics calculated on the remaining 50,000. These very long and time-consuming simulations allowed us to assess community stability after the elapse of the transient dynamics, an aspect of the simulation of this model overlooked by Lehman and Tilman (2000). For each of the combinations of N and γ values, 11 independent simulations were conducted and the results were averaged. Both the Heun integrator and the spectral mimicry method where implemented in the Matlab environment.

In keeping with previous studies we adopt a definition of stability based on the temporal variance of population and community biomass. We measure population (*PS*) and community (*CS*) stability as the inverse of the coefficient of variation (Lehman and Tilman 2000):

$$PS = \frac{\sum_{i=1}^{N} \frac{\mathrm{E_t}[B_i]}{\sqrt{\mathrm{Var_t} B_i}}}{N}$$

$$CS = \frac{\mathrm{E_t}\left[\sum_{i=1}^{N} B_i\right]}{\sqrt{\mathrm{Var_t}\left[\sum_{i=1}^{N} B_i\right]}} \tag{8.4}$$

where $\mathrm{E_t}$ and $\mathrm{Var_t}$ stand for the time average and variance of the argument variable. Furthermore, because previous analyses have identified community evenness as an important factor influencing community stability (Hughes and Roughgarden 1998; Ives and Hughes 2002), we also examine how community evenness (*CE*), calculated at each instant in time and averaged over each of the replicate simulations, i.e.

$$CE = \mathrm{E_t}\left[-\frac{\sum_{i=1}^{N} p_i \ln(p_i)}{\ln(N)}\right], \quad p_i = \frac{B_i}{\sum_{i=1}^{N} B_i}, \tag{8.5}$$

is altered as a function of N and γ.

8.4 RESULTS

8.4.1 Without immigration ($I_i = 0$)

For a given level of diversity, increasing γ (the relative dominance of low frequencies in the environment) has a large qualitative effect on both population and community dynamics (Figure 8.1). For low values of γ we obtain a single steady state defined by sustained population fluctuations of small amplitude for all species (Figure 8.1 top panels). An increase in γ results in "outbreak" dynamics in which each species increases from rare and dominates the community for relatively brief periods of time (Figure 8.1 middle panels). For $\gamma > 1.0$ we obtain metastable dynamics defined by distinct switches between long steady-states with alternating species dominance (Figure 8.1 bottom panels).

For $\beta = 0$, increasing N reduced mean population biomass (Figure 8.2A) but, because of compensatory population dynamics, it had a weak positive effect on total community biomass that saturated quickly for $N > 2$ (Figure 8.2E). For $\beta = 0.2$, whilst the effects on population biomass are unchanged (Figure 8.2B), community biomass (Figure 8.2F) peaked for $N = 4$ and then declined as N is increased from 2 to 24.

For $\beta = 0$, increasing γ from 0 to 2 reduced population stability eight–fold (Figure 8.2C) and community stability 100-fold (Figure 8.2G). This destabilizing effect of γ was associated with strong alterations in the symmetry of community structure as reflected by the declines in community evenness for $\gamma > 0.2$ (Figure 8.3A). For all levels of γ increasing diversity stabilized fluctuations in community biomass, although this stabilizing effect was small (twofold) compared to the destabilizing (100-fold) effect of the environment described above. Furthermore, the stabilizing effect of diversity saturated at $N > 3$ (Figure 8.2G). Similar effects of increasing γ on population (Figure 8.2D) and community stability (Figure 8.2H) were observed for $\beta = 0.2$. However, for $\gamma > 0.2$ and $\beta = 0.2$ communities were destabilized by increasing N (Figure 8.2H). In general, communities were less stable for $\beta = 0.2$ than $\beta = 0$. In sum, in the absence of immigration, irrespective of the value of β, the destabilizing effects of increasing γ outweighed the stabilizing (or destabilizing) effects of increasing N.

8.4.2 With immigration ($I_i = 9.89\text{E} - 4$)

For $\beta = 0$, immigration strongly and qualitatively alters the results for community stability (Figure 8.4G), whilst population stability remains unaltered (Figure 8.4C). The continuous loss of stability described above with increasing γ now only occurs for $N = 2$. For $N > 2$, although community stability still declines 100-fold up to $\gamma = 1.0$, for $\gamma > 1.0$ we obtain a rapid increase in stability such that for $\gamma = 2.0$ the community is more stable than for $\gamma = 0$. Furthermore for $\gamma > 1.5$ there is a strongly stabilizing (100-fold) effect of increasing N from 2 to 3 species, an effect that is

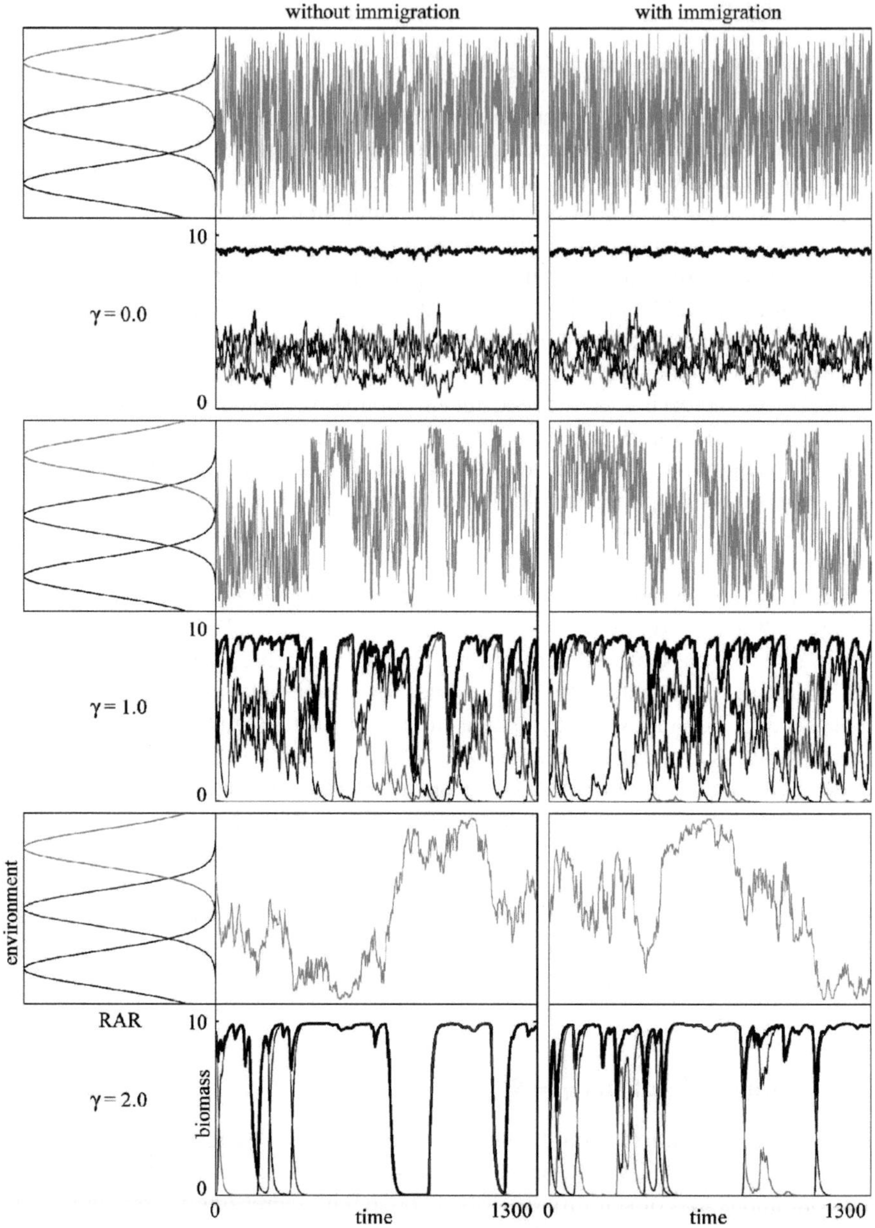

Figure 8.1. Example time series, for the two cases with and without immigration, for the case where $\beta = 0$, and for three levels of $\gamma = 0.0$, 1.0, and 2.0. Environmental fluctuations are shown in grey. Adjacent to these is shown the niche of each species along the environmental gradient (RAR = resource assimilation rate). The mixed grey lines show the fluctuations in species biomass; the black line depicts the fluctuations in total community biomass.

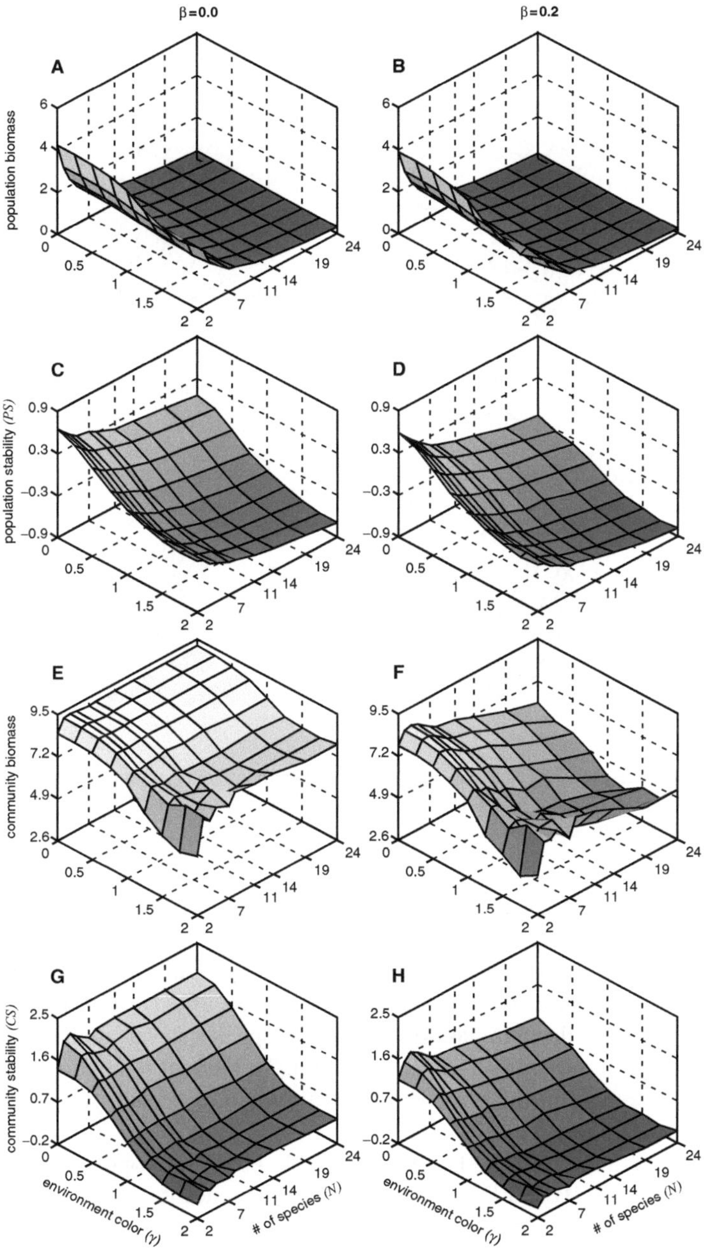

Figure 8.2. Surface plots, for the case without immigration, of mean population (a and b) and community (e and f) biomass and $\log_{10}$ of population (*PS*, c, and d) and community (*CS*, g, and h) stability, as a function of species richness (N) and fluctuation spectra (γ), for the two cases of $\beta = 0$ and $\beta = 0.2$, see text for calculation.

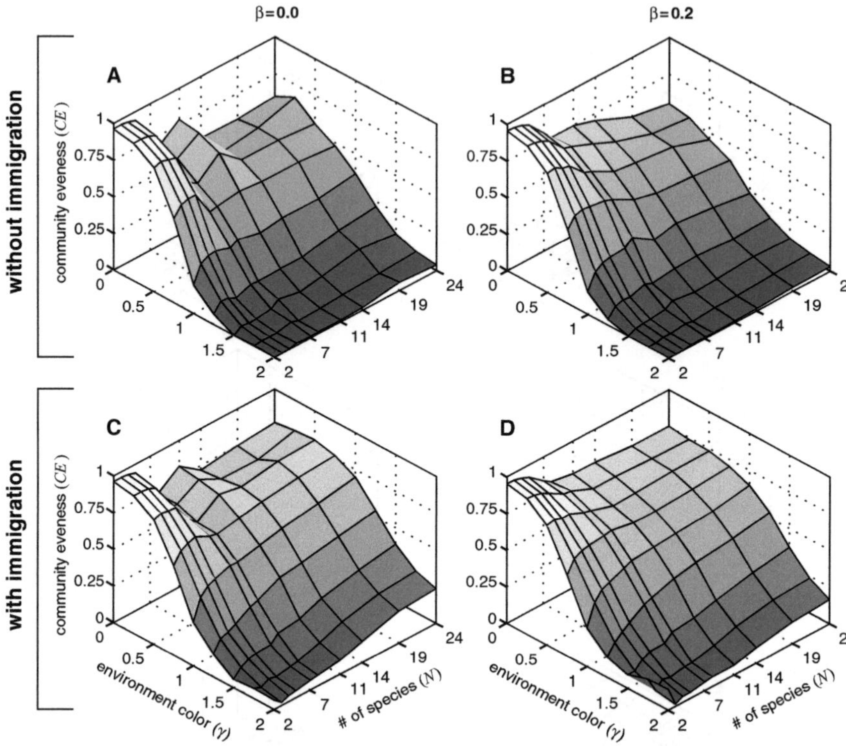

Figure 8.3. Surface plots, for the two cases with and without immigration, of mean instantaneous community evenness *CE* as a function of (N) and fluctuation spectra (β), for the two cases of $\gamma = 0$ and $\gamma = 0.2$, see text for calculation.

substantially greater than the weaker approximately threefold stabilizing effect of increasing N over the same range for $\gamma < 1.5$.

Immigration has a strong effect on community biomass (Figure 8.4E). The presence of a continuous flow of immigrants (a mass effect) from outside the local community ensures that biomass of any single species never attains the very low levels observed in the absence of immigration. This effect is confirmed by greater community evenness in the presence of immigration, especially for $N > 10$ (Figure 8.3C).

Increasing β from 0 to 0.2 had no noticeable effect on population biomass and stability (Figures 8.4B and D). However, we found a negative effect on community biomass and stability (Figures 8.4F and H) for $0 < \gamma < 1.0$ as N was increased. This effect is no longer apparent for $\gamma > 1.0$ and $N = 4$, resulting in the attenuation of the U-shaped relation between γ and community stability.

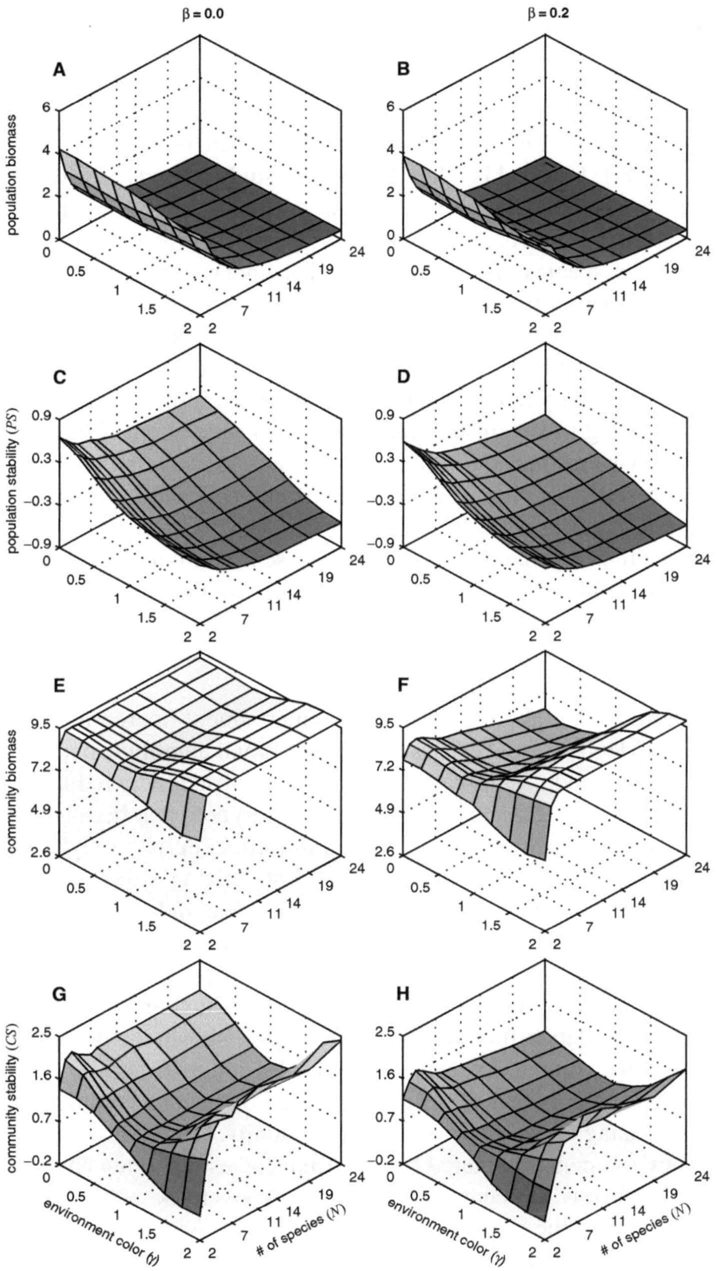

Figure 8.4. Surface plots, for the case with immigration, of mean population (a and b) and community (e and f) biomass and $\log_{10}$ of population (*PS*, c, and d) and community (*CS*, g, and h) stability, as a function of species richness (N) and fluctuation spectra (γ), for the two cases of $\beta = 0$ and $\beta = 0.2$, see text for calculation.

8.5 DISCUSSION

The insurance effect is now considered to be the principal mechanism by which diversity may beget stability (McNaughton 1977; Doak et al. 1998; Hughes and Roughgarden 1998; Ives et al. 1999; Yachi and Loreau 1999; Chesson et al. 2001; Norberg et al. 2001; Ives and Hughes 2002). This study extends previous work using this resource competition model (Lehman and Tilman 2000; Chesson et al. 2001) where the stabilizing effects are a direct consequence of the temporal niche differentiation between species. We have shown that under certain realistic patterns of environmental variation the destabilizing effects of the environment were up to 50 times greater than the stabilizing effect of increasing diversity. These results suggest that the temporal structure of the environment may significantly modulate the stabilizing effects of diversity in a manner that has not been fully appreciated to date.

An understanding of our results requires the consideration of three factors: the width of the environmental tolerance (w_f) of each species, the resource assimilation rate (r_i) of each species, and the scale of autocorrelation in the environment (γ). Because w_f and r_i were held constant across species, it is possible to identify the direct effects of increasing γ in the presence and absence of immigration. Increasing γ destabilized total community biomass for three reasons: (1) because it increased population variability, (2) reduced community evenness, and (3) lengthened the interval (lowered the frequency) between compensatory switches in dominance. We consider these effects in turn below.

Increasing the temporal correlation of the environment (increasing γ) is destabilizing in this model because it causes a shift from small amplitude population fluctuations to large amplitude "outbreak" dynamics that substantially reduced population stability (Figures 8.1, 8.2C and D). It is important to note that because the environmental mean and variance was held constant this effect derives only from a change in the autocorrelation of the environment. A destabilizing effect of autocorrelated environmental fluctuations was also reported by Steele and Henderson (1984) and Gonzalez and Holt (2002), although these studies only considered single species populations. Previous theoretical studies (e.g. Roughgarden 1975; Ripa and Ives 2003) have demonstrated that autocorrelated environmental fluctuations can increase the variance of population fluctuations but none have shown the destabilizing effect this can have at the community level (Figures 8.2G and 8.4G and H).

The second result of note is the effect of the environment on community evenness, and the destabilizing affect this has on the community. Low levels of γ (< 0.2) characterize an environment dominated by high frequencies that often changes state and remains within the environmental tolerance of any given species for relatively brief periods of time. Here the short-term fluctuations of the environment cause community biomass to be distributed relatively evenly across several species (high evenness) at any one point in time (Figures 8.1 and 8.3), total biomass never reaches low levels and community stability is maximal. This scenario corresponds most closely to that explored by recent theoretical analyses that have used linear models to study the effects of diversity on community stability (e.g. Hughes and Roughgarden 1998; Ives and Hughes 2002). This previous work assumed that the variance of the population and community fluctuations is small and community evenness is high (Ives and Hughes 2002), and it is useful that we recover this result for low values of γ (< 0.2). However, by studying the response of a nonlinear model to large and autocorrelated environmental fluctuations we obtained additional results that enrich earlier conclusions. We found that increasing the autocorrelation length of the environment creates large amplitude population and community fluctuations and a low evenness in the biomass distribution across the community. Low evenness precludes the contribution of much of the community to any insurance effect of diversity and this effect of the environment is strongly destabilizing. Ives and Hughes (2002) indicate that low evenness is destabilizing and suggested this might come about by asymmetric interspecific competition. We conclude that autocorrelation in the environment may also be sufficient to create strong unevenness and low stability (see also Gonzalez and Descamps-Julien 2004).

An important feature of autocorrelated environmental fluctuations is the presence of relatively infrequent but abrupt shifts in environmental state (Roy et al. 1996; Alley et al. 2003). Here, this environmental pattern induced a dynamical behavior known as metastability (Galves et al. 1987) in the nonequilibrium dynamics of the community. Metastable dynamics are commonly found in nonlinear systems forced by random variation (Horsthemke and Lefever 1984), and are characterized by the long-term persistence of unstable equilibria and abrupt shifts between unstable equilibria (Galves et al. 1987). In this model, for $\gamma > 1.0$, the abrupt shifts in community biomass are associated with species turnover where the dominant species was almost entirely replaced, typically with a delay (results not shown), by the species optimally adapted to the new prevailing environmental state. Examples of this type of community metastability followed by abrupt reorganization in response to environmental fluctuations can be found in the paleoecology literature, and it has been reported for assemblages as disparate as temperate forests and grasslands, and oceanic foraminifera (e.g. Davis 1986; Tsedakis 1993; Roy et al. 1996; Cannariato et al. 1999).

The major consequence of metastability and abrupt species turnover through time is the strong destabilizing effect at the community level; immigration buffered this effect. Autocorrelated environmental variability resulted in low community evenness that increased the duration of periods of rarity (Steele and Henderson 1984; Caswell and Cohen 1995). In our model immigration stabilized the community in strongly autocorrelated environments ($\gamma > 1.0$) by shortening the recurrence time of rare species, a continuous flow of immigrants maintained the biomass levels of the rare species at a level that allowed rapid recovery as the environment changed state. This result supports the finding of Loreau et al. (2003) that present a similar stabilizing effect of dispersal in source-sink metacommunities. Our results suggest that metastable dynamics may be a common feature of nonequilibrium community dynamics that warrant further study.

Our model, like any other, makes several simplifying assumptions. Clearly, rates of biomass production are not equal across all species in a guild or community, and species are not spaced equally in environmental niche space. The effects of relaxing these assumptions will be reported elsewhere (Gonzalez and DeFeo, unpublished). Furthermore, at very long timescales one might expect to see evolutionary change in the environmental tolerance of component species. With regard to this, Bennett (1997) has argued that millennial-scale climate changes obliterate adaptation to local environments at shorter timescales and that the position and volume of a species' fundamental niche are constrained phylogenetically. In addition, we employed a resource competition model with a single trophic level because recent experiments that employed the same model of resource competition indicate that fluctuation-dependent mechanisms of coexistence may underlie the stabilizing effect of diversity (Descamps-Julien and Gonzalez 2005). Of course all species are embedded in a food web which may alter the results we obtained here. However, Ripa and Ives (Chapter 6, this volume) also report destabilizing effects of environmental autocorrelation on simple food webs using multivariate autoregressive models of multispecies communities. Our results are relevant to recent experimental tests of diversity-stability theory that have demonstrated a destabilizing effect of increasing species richness (Petchey et al. 2002; Gonzalez and Descamps-Julien 2004). For example, Gonzalez and Descamps-Julien (2004) found that increasing species richness destabilized population growth rates. In our model this result was obtained for positive values of β that caused a monotonic decline in the width of a species' niche with increasing species diversity, i.e. a decline in niche niche (w_f) with species addition. Mean community biomass and stability declined under this scenario because the total amount of environmental range covered by the community increased less than linearly with the addition of new species. There is little in the literature on the direct effect of interspecific competition on species niche breadth, thus further experimentation is now needed to establish whether this kind of mechanism is sufficient to explain the experimental results cited above.

These results also raise two issues of current conservation concern. Firstly, human-induced climate change may involve not only changes in mean conditions, but also increased autocorrelation of the environment (Wigley et al. 1998). The effect of such change has been ignored in the ecological literature as a potential climate change impact. Our results suggest that increasing environmental autocorrelation may have a strong destabilizing effect on community stability. Secondly, this effect will be compounded by the loss of rare species when the environment is not at their competitive optima (e.g. Jackson and Weng 1999). Palynological evidence suggests that species maladapted to prevailing environmental conditions have persisted in the past at very low densities in regionally dispersed local refugia (Schauffler and Jacobson 2002). Following environmental change these species can spread rapidly and come to dominate the regional community in a matter of generations (Gear and Huntley 1991; Solomon and Kirilenko 1997). Habitat fragmentation is currently threatening the persistence and dispersal capacity of many rare species and hence their likelihood of contributing to community stability. The results of our model suggest that the interactive effects of climate change (increasing autocorrelation) and the loss of rare species may have hitherto unforeseen consequences for future community stability.

8.6 ACKNOWLEDGMENTS

We would like to thank Kevin McCann and David Vasseur for inviting us to contribute to this volume in honor of Peter Yodzis' contribution to the field of theoretical ecology; his work is a great inspiration to us both. Andrew Gonzalez acknowledges the financial support of the *Quantitative Ecology* Coordinated Action Incentive of the Ministry of Research (France), the Canada Research Chair Program and NSERC. Oscar De Feo acknowledges the financial support of the Science Foundation Ireland. We thank Pablo Inchausti, Michel Loreau, Per Lundberg, Owen Petchey, and Brian McGill for their comments on an earlier version.

8.7 LITERATURE CITED

Allen, J.R.M., Brandt, U., Brauer, A. et al. 1999. Rapid environmental changes in southern Europe during the last glacial period. *Nature* 400: 740–743.

Alley, R.B., Marotzke, J., Nordhaus, W.D. et al. 2003. Abrupt climate change. *Science* 299: 2005–2010.

Austin, M.P., Nicholls, A.O., and Margules, C.R. 1990. Measurement of the realized qualitative niche: environmental niches of five eucalyptus species. *Ecological Monographs* 60: 164–177.

Bennett, K.D. 1997. *Evolution and ecology: the pace of life*. Cambridge University Press, Cambridge.

Bengtsson, J. 1994. Temporal stability in forest soil communities. *Journal of Animal Ecology* 63: 653–665.

Braswell, B.H., Schimel, D.S., Linder, E. et al. 1997. The response of global terrestrial ecosystems to interannual temperature variability. *Science* 278: 870–872.

Cannariato, K.G., J.P. Kennett, and R.J. Behl. 1999. Biotic response to late quaternary rapid climate switches in Santa Barbara Basin: ecological and evolutionary implications. *Geology* 27: 63–66.

Caswell, H. and J.E. Cohen. 1995. Red, white and blue: environmental variance spectra and coexistence in metapopulations. *Journal of Theoretical Biology* 176: 301–316.

Chesson, P. 2000. Mechanisms of maintenance of species diversity. *Annual Review of Ecology and Systematics* 31: 343–366.

Chesson, P., S.W. Pacala, and C. Neuhauser, C. 2001. Environmental niches and ecosystem functioning. In: *The functional consequences of biodiversity* (eds., A.P. Kinzig, S.W. Pacala, and D. Tilman), pp. 213–245. Princeton University Press, Princeton.

Cohen, J.E., C.M. Newman, A.E. Cohen et al. 1999. Spectral mimicry: a method of synthesizing matching time series with different Fourier spectra. *Circuits Systems Signal Process* 18: 431–442.

Cuddington, K.M. and P. Yodzis. 1999. Black noise and population risk. *Proceedings of the Royal Society of London, Series B* 269: 969–973.

Descamps-Julien, B. and A. Gonzalez. 2005. Stable coexistence in a fluctuating environment: an experimental demonstration. *Ecology* 86: 2815–2824.

Davis, M.B. 1986. Climate instability, time lags, and community disequilibrium. In: *Community ecology* (eds., J. Diamond. and T.J. Case), pp. 269–284. Harper & Row, New York.

Doak, D.F., D. Bigger, E.K. Harding, et al. 1998. The statistical inevitability of stability–diversity relationships in community ecology. *American Naturalist* 151: 264–276.

Dynesius, M. and R. Jansson. 2000. Evolutionary consequences of changes in species' geographical distributions driven by Milankovitch climate oscillations. *Proceedings of the National Academy of Sciences of the USA* 97: 9115–9120.

Francis, R.C., S.R. Hare, A.B. Hollowed, et al. 1998. Effects of interdecadal climate variability on the oceanic ecosystems of the NE Pacific. *Fisheries Oceanography* 7: 1–21.

Galves, A., E. Olivieri, and M.E. Vares. 1987. Metastability for a class of dynamical systems subject to small random perturbations. *The Annals of Probability* 15: 1288–1305.

Gear, A.J. and B. Huntley 1991. Rapid changes in the range limits of Scots pine 4000 years ago. *Science* 251: 544–547.

Gonzalez, A. and B. Descamps-Julien. 2004. Population and community stability in a fluctuating environment. *Oikos* 106: 105–116.

Gonzalez, A. and R.D. Holt. 2002. The inflationary effects of environmental fluctuations in source-sink systems. *Proceedings of the National Academy of Sciences of the USA* 99: 14872–14877.

Halley, J.M. 1996. Ecology, evolution and $1/f$ noise. *Trends in Ecology and Evolution* 11: 33–37.

Halley, J.M. and W.E. Kunin, 1999. Extinction risk and the $1/f$ family of noise models. *Theoretical Population Biology* 56: 215–230.

Heino, M. 1998. Noise colour, synchrony and extinctions in spatially structured populations. *Oikos* 83: 368–375.

Holt, R.D., Barfield, M., and A. Gonzalez. 2003. Impacts of environmental variability in open populations and communities: 'inflation' in sink environments. *Theoretical Population Biology* 64: 315–331.

Horsthemke, W. and R. Lefever. 1984. *Noise induced transitions: theory and applications in physics, chemistry, and biology*. Springer, Berlin.

Hughes, J.B. and J. Roughgarden. 1998. Aggregate community properties and the strength of community interactions. *Proceedings of the National Academy of Sciences of the USA* 95: 6837–6842.

Hulme, M., E.M. Barrow, N.W.P.A. Arnell, et al. 1999. Relative impacts of human-induced climate change and natural climate variability. *Nature* 397: 688–691.

Inchausti, P. and J. Halley. 2002. Investigating long-term ecological variability using the Global Population Dynamics database. *Science* 293: 655–657.

Ives, A.R., K. Gross, and J.L. Klug. 1999. Stability and variability in competitive communities. *Science* 286: 542–544.

Ives, A.R. and J.B. Hughes. 2002. General relationships between species diversity and stability in competitive systems. *American Naturalist* 159: 388–395.

Jackson, S.T. and C. Weng. 1999. Late quaternary extinction of a tree species in eastern North America. *Proceedings of the National Academy of Sciences of the USA* 96: 13847–13852.

Knapp, A.K., F.A., Fay, J.M., Blair, et al. 2002. Rainfall variability, carbon cycling, and plant species diversity in a mesic grassland. *Science* 298: 2202–2205.

Lehman, C.L. and D. Tilman. 2000. Biodiversity, stability, and productivity in competitive communities. *American Naturalist* 156: 534–552.

Levin, S.A. 1992. The problem of pattern and scale in ecology: The Robert H. MacArthur award lecture. *Ecology* 73: 1943–1967.

Loreau, M., N. Mouquet, and A. Gonzalez. 2003. Biodiversity as spatial insurance in heterogeneous landscapes. *Proceedings of the National Academy of Sciences of the USA* 100: 12765–12770.

Mandelbrot, B.B. and J.R. Wallis. 1969. Some long-run properties of geophysical records. *Water Resources and Research* 5: 321–339.

McGowan, J.A., D.R. Cayan, and L.M. Dorman. 1998. Climate-ocean variability and ecosystem response in the northeast Pacific. *Science* 281: 210–217.

McNaughton, S.J. 1977. Diversity and stability of ecological communities: a comment on the role of empiricism in ecology. *American Naturalist* 111: 515–525.

Morales, J.M. 1999. Viability in a pink environment: why "white noise" models can be dangerous. *Ecology Letters* 2: 228–232.

Norberg, J., D. Swaney, P. Dushoff, J., et al. 2001. Phenotypic diversity and ecosystem functioning in changing environments: A theoretical framework. *Proceedings of the National Academy of Sciences of the USA* 98: 11376–11381.

Pelletier, J.D. 2002. Natural variability of atmospheric temperatures and geomagnetic intensity over a wide range of time scales. *Proceedings of the National Academy of Sciences of the USA* 99: 2546–2553.

Petchey, O.L. 2000. Environmental colour affets the dynamics of single species populations. *Proceedings of the Royal Society of London, Series B* 267: 747–754.

Petchey, O.L., A. Gonzalez, and H.B. Wilson. 1997. Effects on population persistence: the interaction between environmental noise colour, intraspecific competition and space. *Proceedings of the Royal Society of London, Series B* 264: 1841–1847.

Petchey, O.L., T. Case, P. Lin Jiang, et al. 2002. Species richness, environmental fluctuations, and temporal change in total community biomass. *Oikos* 99: 231–240.

Quarteroni, A., R. Sacco, and F. Saleri. 1999. *Numerical mathematics.* Springer, Berlin.

Rahel, F.J. 1990. The hierarchical nature of community persistence: a problem of scale. *American Naturalist* 136: 328–344.

Ripa, J. and A.R. Ives. 2003. Food web dynamics in correlated and autocorrelated environments. *Theoretical Population Biology* 64: 369–384.

Ripa, J. and P. Lundberg. 1996. Noise colour and the risk of population extinctions. *Proceedings of the Royal Society of London, Series B* 263: 1751–1753.

Ripa, J., P. Lundberg, and V. Kaitala. 1998. A general theory of environmental noise in ecological food webs. *American Naturalist* 151: 256–263.

Roughgarden, J. 1975. A simple model for population dynamics in stochastic environments. *American Naturalist* 109: 713–736.

Roy, K., J.W. Valentine, D. Jablonski, et al. 1996. Scales of climatic variability and time averaging in Pleistocene biotas: implications for ecology and evolution. *Trends in Ecology and Evolution* 11: 458–463.

Solomon, A.M. and A.P. Kirilenko, 1997. Climate change and terrestrial biomass: what if trees do not migrate? *Global Ecology and Biogeography Letters* 6: 139–148.

Schauffler, M. and G.L. Jacobson Jr. 2002. Persistence of coastal spruce refugia during the Holocene in northern New England, USA, detected by stand-scale pollen stratigraphies. *Journal of Ecology* 90: 235–250.

Schmida, A. and M.V. Wilson. 1985. Biological determinants of species diversity. *Journal of Biogeography* 12: 1–20.

Shugart, H.H. 1998. *Terrestrial ecosystems in changing environment.* Cambridge University Press, Cambridge.

Steele, J.H. 1985. A comparison of terrestrial and marine ecological systems. *Nature* 313: 355–358.

Steele, J.H. and E.W. Henderson. 1984. Modeling long-term fluctuations in fish stocks. *Science* 224: 985–987.

Stenseth, N.C., A. Mysterud, G. Otterson, et al. 2002. Ecological effects of climate fluctuations. *Science* 297: 1292–1296.

Tsedakis, P.C. 1993. Long-term tree populations in northwest Greece through multiple Quaternary climate cycles. *Nature* 364: 437–440.

Vasseur, D.A. and P. Yodzis. 2004. The color of environmental noise. *Ecology* 85: 1146–1152.

Venrick, E.L. 1990. Phytoplankton in an oligotrophic ocean: species structure and interannual variability. *Ecology* 71: 1547–1563.

Whitlock, C. and P.J. Bartlein. 1997. Vegetation and climate change in northwest America during the past 125kyr. *Nature* 388: 57–61.

Wichmann, M.C., K. Johst, K.A. Moloney, C. Wissel, and F. Jeltsch. 2003. Extinction risk in periodically fluctuating environments. *Ecological Modelling* 167: 221–231.

Wigley, T.M.L., R.L. Smith, and B.D. Santer. 1998. Anthropogenic influence on the autocorrelation structure of hemispheric-mean temperatures. *Science* 282: 1676–1679.

Willis, K.J., A. Kleczkowski, and S.J. Crowhurst. 1995. 124,000-year periodicity in terrestrial vegetation change during the late Plioene. *Nature* 397: 685–688.
Yachi, S. and M. Loreau. 1999. Biodiversity and ecosystem productivity in a fluctuating environment: the insurance hypothesis. *Proceedings of the National Academy of Sciences of the USA* 96: 1463–1468.

APPENDIX 8A

Normalization: The resource competition model where each species has its optimal competitive ability at some value of the environmental factor is,

$$\begin{cases} \dot{R} = \alpha(S-R) - \sum_{i=1}^{N} g_i(x)\frac{Q_i R B_i}{R+K_i} \\ \dot{B}_i = B_i\left[g_i(x)\frac{R}{R+K_i} - m_i\right] \\ g_i = r_i \exp\left[-\frac{1}{2}\left(\frac{x-\tau_i}{w}\right)^2\right] \end{cases} \quad i \in \{1, 2, \ldots, N\} \tag{8A.1}$$

This model contains $5N + 3$ parameters which do not act independently on the dynamics of the system. Rescaling the state variables and time as:

$$R \to \frac{R}{S}, \quad B_i \to \frac{B_i Q_i}{S}, \quad t \to \alpha t, \tag{8A.2}$$

we obtain,

$$\begin{cases} \dot{R} = 1 - R - \sum_{i=1}^{N} g_i(x)\frac{R B_i}{R+K_i} \\ \dot{B}_i = B_i\left[g_i(x)\frac{R}{R+K_i} - m_i\right] \\ gi = r_i \exp\left[-\frac{1}{2}\left(\frac{x-\tau_i}{w}\right)^2\right] \end{cases} \quad i \in \{1, 2, \ldots, N\} \tag{8A.3}$$

where the new parameters are rescaled as,

$$r_i \to \frac{r_i}{\alpha}, \quad K_i \to \frac{K_i}{S}, \quad m_i \to \frac{m_i}{\alpha}, \tag{8A.4}$$

and the new model has $4N + 1$ independent parameters.

We can also normalize the environmental variable, by supposing x is given by, $x = \bar{x} + \Delta x$, where $\bar{x}$ is the mean value and Δx are the environmental fluctuations which are within the range $[\Delta x_m, \Delta x_M]$. We rescale the variable x, the parameters τ_i, and w with respect to $\Delta x_M - \Delta x_m$, so that x varies between –1 and +1 and the width of the environmental tolerance is,

$$w \to \frac{w}{\Delta x_M - \Delta x_m} \tag{8A.5}$$

We assume τ_i evenly spread across the range of environmental fluctuations, and

$$w = \frac{w_f}{N^\beta} \tag{8A.6}$$

where, w is equivalent to the realized niche width and w_f the fundamental niche. Thus for a $\beta = 0$ we obtain the case typical of the standard model, whilst for $\beta = 0.2$ the width of the environmental tolerance declines monotonically with increasing species diversity.

Finally, we have modified the species i biomass equation as follows:

$$\dot{B}_i = B_i \left[g_i(x) \frac{R}{R + K_i} - m_i \right] + I_i \tag{8A.7}$$

to introduce the effects of a continuous flow (I_i) of exogenous immigrants (a mass effect) from outside the local community. To give a relative measure to the immigration flow, we can set $I_i = p_i \bar{B}_{i,M}$ where $\bar{B}_{i,M}$ is the maximal species density achievable by species i and p_i is a percent measure of the immigration flow with respect to $\bar{B}_{i,M}$. Hence,

$$I_i = \frac{p_i \left(r_i - m_i + p_i - K_i \left(m_i - p_i\right)\right)}{\left(m_i - p_i\right)\left(r_i - m_i + p_i\right)} \tag{8A.8}$$

which gives $I_i = 9.8989\text{E} - 4$ for the parameters at the values mentioned in the text ($m_i = 0.1$, $K_i = 0.1$, and $r_i = 1.0$) and $p_i = 0.01\%$.

CHAPTER 9

EFFECTS OF ENVIRONMENTAL VARIABILITY ON ECOLOGICAL COMMUNITIES: TESTING THE INSURANCE HYPOTHESIS OF BIODIVERSITY IN AQUATIC MICROCOSMS

OWEN L. PETCHEY

Department of Animal and Plant Sciences, University of Sheffield, Sheffield, S10 2TN
Phone: + 44 0114 2220029, Fax: + 44 011 4 2220002, E-mail: O.Petchey@Sheffield.ac.uk

9.1 ABSTRACT

The insurance hypothesis states that communities with greater numbers of species will be more stable than communities with fewer species. Various theoretical realisations of this hypothesis and some empirical evidence have been interpreted as general support for the hypothesis. Here, I suggest that very few studies can claim to accurately test the hypothesis. In particular, this is because few studies are sufficiently long in duration, and because sometimes community level variability is measured inappropriately. There are very few experiments (perhaps only six) that meet the assumptions of the theoretical models closely. Across these, around a half of all tests indicate support for the insurance hypothesis, and half show no support. This is probably too small a number of experiments to draw any solid conclusions and should encourage more experimental and observational tests. In particular, there have been

David A. Vasseur and Kevin S. McCann, The Impact of Environmental Variability on Ecological Systems, 179–196.

very few investigations of how environmental variability interacts with diversity to determine community stability. Other directions for future research include replacing species richness with diversity of responses to environmental change, and developing a more mechanistic framework for understanding empirical results.

9.2 INTRODUCTION

The relationship between complexity and stability has interested ecologists for many decades (MacArthur 1955; Elton 1958; Gardner and Ashby 1970; May 1972; McCann 2000). One reason for this may be the practical consequences of knowing, for example, that depauparate communities suffer greater risk of disease outbreak or greater risk of invasion by alien species. Another reason may be the variety of meanings of complexity and stability. Stability can be, for example, how much an ecological entity (population, community, ecosystem) resists a disturbance (resistance), how quickly it recovers from the effects of a disturbance (resilience), or how much it varies through time in the face of continued disturbances (temporal variability). Complexity can include the number of species in a community, the number of interactions between those species, and the diversity of trophic roles played by species. Consequently, there are many questions about relationships between the two. Indeed, Pimm (1984) identified 20 different questions about complexity and stability and suggested that this catalogue was not exhaustive. A more recent review increased the number of components of stability to seven, which, if used in Pimm's framework, would result in 42 different questions about complexity and stability (Loreau et al. 2002).

The complexity—stability question addressed here is motivated by the possible consequences of species loss for the services that ecosystems provide humanity (Daily 1997). These services are often derived from processes that occur at the ecosystem level of ecological organisation (e.g. as opposed to population level). Consequently, ecologists are asking how species richness (the measure of complexity) might affect the magnitude and variability of ecosystem processes (Cottingham et al. 2001; Hooper et al. 2005). While some studies have accomplished this by direct measurement of an ecosystem process rate, many use the total standing stock of biomass as a surrogate for ecosystem process rates. Because total biomass is the sum of the biomass of the individual populations in a community, it is known as an "aggregate property" of ecological communities. Under some circumstances it is relatively clear that loss of species will have a negative effect on the magnitude of some ecosystem processes (and aggregate properties). For example, a majority of experiments conducted on single trophic level communities in grasslands suggest that less speciose communities produce less biomass and leach more nutrients than communities with more species (Hector et al. 1999; Tilman et al. 2001; Roscher et al. 2004). Such effects are far from general, however. Spatial and temporal scale, trophic complexity, ecosystem type, and sometimes different analytical methods can each have important effects on the functional consequences of changes in species richness (Wardle 2002; Hooper et al. 2005).

There is some suggestion, however, that the consequences of reduced species richness might have a more general effect on the temporal variability of ecosystem processes (Cottingham et al. 2001; Loreau et al. 2002; Worm and Duffy 2003). Specifically, that ecosystems with more species will have less variable (more stable) dynamics of productivity, nutrient flux, and decomposition. This potential stabilising effect of species richness on the temporal variability of ecosystem properties may have its origins in MacArthur's (1955) work and was formalised more by McNaughton (McNaughton 1977). Several influential studies revived interest in the subject (e.g. Tilman 1996; McGrady-Steed et al. 1997; Naeem and Li 1997) and it has subsequently been the subject of much theoretical and experimental investigation and is the focus of this article. The "insurance hypothesis" (its given name) is that increases in species richness will reduce community and ecosystem level variability. Here, I present a review of empirical evidence of effects of species richness on the temporal variability of populations, communities, and ecosystems. The next section details some important methodological considerations that limit the number of experimental studies that provide clear and direct tests of the insurance hypothesis. Subsequent sections discuss the results of these few experiments and their ecological significance.

9.3 METHODOLOGICAL CONSIDERATIONS

Most theory about the insurance hypothesis concerns long-term dynamics, where individuals have the opportunity to reproduce for many generations, interspecific interactions influence population size through consumption and competition, and transient dynamics caused by initial conditions are of limited importance. More specifically, stabilising effect of richness rely on the potential for random (Doak et al. 1998; Tilman et al. 1998) or compensatory population growth (Ives et al. 1999) and/or for population sizes to respond to changes in environmental conditions (Ives et al. 1999; Yachi and Loreau 1999). Compensatory growth can occur at the individual level and this may be an important mechanism in short-term experiments with plant species (Pfisterer and Schmid 2002). A strong test of the insurance hypothesis requires an experiment long enough in duration to include many generations of the study organisms and consequently all these biological processes (Connell and Sousa 1983). Therefore multigenerational experiments closely match the assumptions of the theoretical models that produce an insurance effect of biodiversity. Another advantage of multigenerational experiments is that they escape the criticism that results are strongly affected by the number of individuals of each species used to assemble communities (Schwartz et al. 2000). In particular, assembly usually involves equal numbers of individuals of each species, whereas natural communities have relatively uneven distribution of individuals among species. In all the experiments discussed below, sufficient population growth and decline occurs that evenness in abundances is not a function of initial conditions.

Some of the most important experimental studies have used model communities containing small bodied organisms (McGrady-Steed et al. 1997; Naeem and Li 1997). These organisms have fast generation times and allow for long-term population dynamics to take place and be recorded in a relatively short amount of time. Their small size also allows for experiments with relatively high levels of replication of very uniform ecological units. Treeholes, rockpools, and moss patches are natural ecosystems that have answered important ecological questions with some success (Srivastava et al. 2004). Artificial microcosms provide another setting for conducting experiments with small organisms and many generations (Jessup et al. 2004). While these lack the realism of natural microcosms, they offer greater control over experimental conditions and community composition and increase the precision of experiments (Morin 1998).

Experiments that aim to quantify a relationship between species richness and ecosystem variability have benefited from research into the design of experiments that investigate relationships between species richness and the magnitude of ecosystem functioning (Allison 1999; Schmid et al. 2002). First, direct manipulation of species richness produces results that are more easily interpreted than if species richness is manipulated indirectly, for example by fertilizer addition (Huston 1997). Easier interpretation occurs because the species richness treatment is not confounded with the environmental driver. Direct manipulation of species richness necessitates choices about (i) what species richness levels to use, (ii) how many compositions to have within each richness level, (iii) how to assign species to compositions, and (iv) how many replicates to have of each composition. The answers to (ii) and (iii) are often at least two and randomly. The resulting experiments are called combinatorial biodiversity experiments (Naeem 2002). They dissociate the species richness treatment from variation in composition. Random assignment of species to compositions, or more commonly semi-random assignment, ensures relatively uniform occurrence of species along the species richness manipulation. Semi-random can include random assignment of species within constraints placed by simple biological rules (predators must have at least one prey) or the constraint that every species should occur once within the compositions in a species richness level (Roscher et al. 2005). All of the experiments discussed in detail below have at least three species richness levels, multiple compositions within richness levels, and random or semi-random assignment of species to compositions (Table 9.1). All but one also include multiple replicates of each composition. If one is only interested in effects of species richness and not composition (Petchey et al. 2002), replicates of compositions are unnecessary and their absence means that experimental resources can be directed towards more richness levels and/or compositions within richness levels.

It is (obviously) essential that multiple observations of the same ecological variable are needed to calculate temporal variability (Table 9.1). The minimum being two (Petchey et al. 2002), while substantially more than this is desirable

Table 9.1. A summary of direct experimental manipulations of species richness that have recorded effects on total biomass of a community or functional groups, or ecosystem process rates (see Figures 9.1 and 9.2)

								Effect of diversity on:			
Study	Diversity levels	Species composition	Trophic complexity	Additional treatments	Length of time series	Response variable	Measure of variability	Temporal variability	Population variability	Total biomass	Summed covariances
(McGrady-Steed and Morin 2000)	3, 5, 10, 15, 31	Random and replicated	Algae, bacteria, bacterivores, herbivores, predators	None	6	Total biomass within four functional groups	Coefficient of variation (CV)	3 or 4 functional groups were stabilised	16 of 24 populations showed no significant effect	2 of 4 functional groups increased	Not presented
(Morin and McGrady-Steed 2004)[a]	0, 3, 5, 10, 15, 20, 25, 31	Random and replicated	Algae, bacteria, bacterivores, herbivores, predators	None	6	CO_2 flux rate	Standard deviation of variable	No effect	Not relevant	Not relevant	Not relevant
(Steiner et al. 2005)	4, 8, 16	Random and replicated	Algae, bacteria, bacterivores, herbivores, predators	Two levels of nutrient supply	7	Total community biomass	Standard deviation of log transformed variable	Decreased	Not presented	Probably increases	No effect

Table 9.1. (Continued)

Study	Diversity levels	Species composition	Trophic complexity	Additional treatments	Length of time series	Response variable	Measure of variability	Effect of diversity on: Temporal variability	Population variability	Total biomass	Summed covariances
(Petchey 2000)	1, 2, or 3 prey species	Random and replicated	Prey (protist) bacterivores) and one predatory protist	None	10	Total prey biomass	CV	No effect	Not presented	Not presented	Not presented
						Predator abundance	CV	Decreased	Not relevant	Not relevant	Not relevant
(Petchey et al. 2002)	2, 4, 6, 8	Random and unreplicated	Bacterivorous protists and small metazoans	Five patterns of temperature fluctuations	2	Total biomass	CV	No effect	Increases	Increases	Decreases (becomes more negative)
						CO_2 flux rate	CV	Increased	Not relevant	Not relevant	Not relevant
(Gonzalez and Descamps-Julien 2004)	1, 3, 6	Random and replicated	Algae (6 species) and one herbivorous rotifer	Three patterns of temperature fluctuations	32	Total algal biomass	CV	No effect	Increases	Increases	Increases
						Herbivore abundance	CV	No effect	Not relevant	Not relevant	Not relevant

(e.g. Gonzalez and Descamps-Julien 2004). Multiple data points can also increase confidence that dynamics are relatively stable (McGrady-Steed and Morin 2000) or allow removal of trends present in the times series (Steiner et al. 2005).

Ecological variability occurs through space as well as time. Misunderstanding about whether questions, studies, and critiques are about variation in space or time (or both simultaneously) has lead to some confusion that needs resolving. Both MacArthur (1955) and McNaughton (1977) began by considering temporal variability – variation in ecological properties that occurs through time within a local community. McNaughton in his first figure presents a:

> simple graphical model of how community diversity may increase temporal stability of community properties

(McNaughton 1977). For contrast, consider the spatial differences in species composition that occur among local communities (beta diversity) that is important in other contexts. Naeem (Naeem 1998) subsequently developed the concept of *ecosystem reliability*, which in the original publication is the continued functioning of a single system (local community) through time (Naeem 1998). Most recent theoretical developments of what has increasingly become known as the insurance hypothesis of biodiversity concern temporal variability and consequently this is what I focus on in this article (Doak et al. 1998; Hughes and Roughgarden 1998; Tilman et al. 1998; Ives et al. 1999; Yachi and Loreau 1999; Hughes and Roughgarden 2000; Gonzalez this volume).

Ironically, two particularly influential experimental studies then measured variability *among* local communities either alone or in combination with temporal variability within local communities (McGrady-Steed et al. 1997; Naeem and Li 1997). In doing so they made it difficult to understand how their results related to the insurance hypothesis (because it only concerns temporal variability). The source of difficulty was that the species richness gradient created for the experiment also created a gradient in compositional similarity among local communities within a richness level (Wardle 1998; Fukami et al. 2001). The less rich communities were compositionally less similar than those containing more species. This creates the expectation of higher variation in ecosystem functioning *among* the depauparate communities compared to among the speciose ones (Fukami et al. 2001). This "similarity hypothesis" of biodiversity was presented as an analogue of the insurance hypothesis (i.e. that it could produce the same qualitative patterns) and some researchers have since mistaken it as applying to studies of temporal variability. As the authors made clear, the similarity hypothesis does not concern temporal variability within a local community (Fukami et al. 2001) and insofar as the insurance hypothesis concerns only this, the two are unrelated. For this reason, and though they are important studies, I will not refer to McGrady-Steed et al. (McGrady-Steed et al. 1997) or Naeem and Li (Naeem and Li 1997), though a further analysis of McGrady-Steed et al. will be discussed (Morin and McGrady-Steed 2004).

There are many methods for estimating the temporal variability of an ecological time series (Gaston and McArdle 1994) and it is critical to choose one that applies well to the ecological hypothesis of interest (Cottingham et al. 2001). Those that match most ecologists intuitive notion of variability are proportional (rather than arithmetic), so that doubling represents the same amount of change regardless of absolute values (Gaston and McArdle 1994). Among these, the standard deviation of log transformed values and the coefficient of variation of untransformed values have reasonably desirable properties. Of the studies discussed here, only one uses a different measure of temporal variability: Morin and McGrady-Steed (2004) use the standard deviation of untransformed values (Table 9.1). Their reasons were (i) that while abundances and biomasses cannot be negative, the measures of gas flux they were interested in were sometimes positive (net production of CO_2) and sometimes negative (net absorption of CO_2), making log transformation impossible; and (ii) the standard deviation and mean of the data were not significantly correlated making division by the mean (which was small) redundant.

The work discussed below includes multi-generation experiments, and, due to the interpretational difficulties associated with observational studies of complexity – stability patterns, also involves manipulations of species richness. Though there have been many other important experiments (Cottingham et al. 2001; Loreau et al. 2002; Romanuk and Kolasa 2002); the relatively few that meet these criteria are listed in Table 9.1.

9.4 EXPERIMENTAL RESULTS

Six experiments are considered (Table 9.1; Figures 9.1 and 9.2); these contain a total of 12 assessments of effects of species richness on community/ecosystem level temporal variability. Eight have the total biomass of a group of species as the response variable and four of these indicated a stabilising effect of species richness, while four indicated no effect. Two of the assessments were of effects of species richness on the temporal variation of an ecosystem level process, CO_2 flux. One of these indicated a destabilising effect of species richness and the other no significant effect. The remaining two assessments were of the effects of resource species richness on the temporal variability of a consumer population. One showed a stabilising effect of resource richness on the consumer population, the other showed no significant effect. In summary, effects of species richness were destabilising in one of twelve cases, stabilising in five, and non-significant in six. This is probably too small a sample to make an strong generalisation about the effects of species richness on temporal variability of community and ecosystem variables.

In the five examples of stabilising effects of species richness, the strength of the stabilizing effect varied. Standardised to the effect of adding five species, the decrease in variability would be by a factor of approximately 0.97 (a 3% reduction in variability) (Steiner et al. 2005), and 0.99, 0.95, 0.98 (McGrady-Steed and Morin 2000). It is difficult to calculate the expected decrease in variability caused by adding five species in (Petchey 2000) because species richness changed by only two species in the study and extrapolation to the effect of five is problematic. Nevertheless, changes in species richness of five species do not appear to have large effects on temporal variability, even when only significant stabilising effects are considered.

9.5 VARIABILITY OF POPULATIONS

Temporal variability of aggregate communities properties such as total community biomass, or the biomass of trophic groups is made up of the variability of biomass of individual populations (Tilman 1996). Similarly, temporal variability in ecosystem properties such as primary production, or gas flux (a proxy for energy use), are linked to variability in the size of the populations in the ecosystem. If follows that understanding population level fluctuations will contribute towards understanding of the determinants of community and ecosystem level variability. Indeed, most of the theoretical models that link species richness and ecosystem variability are built around the dynamics of populations (Doak et al. 1998; Yachi and Loreau 1999) and aid understanding of community level variability by highlighting the factors that translate population fluctuations into community and ecosystem level fluctuations. There are two general methods taken, the first is rather phenomenological, the second more mechanistic.

If the temporal change in an aggregate community property is measured by the coefficient of variation then

$$\text{Temporal change} = \frac{\text{SD(Total biomass)}}{\text{Total biomass}} = \frac{\sqrt{\sum_{i}^{s} \mathrm{Var}(B_i) + \sum_{i}^{s} \sum_{i \neq j}^{s} Cov(B_i, B_j)}}{\sum_{i=1}^{s} \overline{B}_i} \tag{9.1}$$

Where $\overline{B}_i = \frac{1}{n} \sum_{t=1}^{n} B_{i,t}$ and $B_{i,t}$ is the biomass of the ith (or jth) of s species during the tth of n samples through time (Lehman and Tilman 2000). This equation provides a tool for investigating how three components of temporal change contribute stabilising and destabilising influences along a species richness gradient. If increases in species richness cause increased population level variability ($\mathrm{Var}(B_i)$) this will tend to destabilise community level variability. However, if increases in richness cause species to fluctuate asynchronously (negative covariance: $\mathrm{Cov}(B_i, B_j)$) there will be a stabilising effect of richness on community variability. The denominator in the equation, the sum of population sizes, suggests that increases in total biomass along the richness gradient will lend a stabilising influence. It is, however, the balance of the three terms that determines how variability changes along a richness gradient. For example, population variability may be destabilised by richness while the community is stabilised (Tilman 1996), perhaps because of increasing asynchrony along the richness gradient.

Only three of the six case studies reported effects of richness on population variability; two reported an increase with increasing richness and one no effect (Table 9.1). Only two of the studies reported covariances between populations, one showed a stabilising effect of richness (covariances became more negative as richness increased) but the other showed no effect (Table 9.1). All of the four studies that reported change in total biomass along the richness gradient found that it increased with increasing species richness (Table 9.1).

Analysing effects of species richness on the statistical components of temporal variability cannot identify the mechanisms that cause changes in total biomass, total variance, and total covariance as a result of changes in diversity. For example, Ives et al. (1999) show that the variability of competitive communities is stabilised by increases in species richness only if species respond differently to environmental fluctuations. This could cause populations to vary asynchronously and create negative covariance. But asynchronous fluctuations could also result from species interactions such as resource competition or predation. Consequently, one cannot conclude that species respond differently to environmental variability just because they vary asynchronously. Similarly, changes in population variability and total biomass cannot easily be attributed to a particular biological mechanism or process. This is why methods such as direct experimental manipulation of putative mechanisms and multivariate autoregressive models (Ives 1995) are more appropriate for disentangling the mechanisms that contribute towards any insurance effect of biodiversity (Cottingham et al. 2001).

9.6 ENVIRONMENTAL VARIABILITY TREATMENTS

A 2001 review of research about the functional consequences of biodiversity wrote:

> Experiments in which both diversity and environmental fluctuations are controlled [manipulated] are now needed to perform rigorous tests of the insurance hypothesis

(Loreau et al. 2001). I have added the word "manipulated" to suggest that diversity and environmental fluctuations be experimental treatments. Four years later this is truer than ever. Only three of the six case studies presented here involved manipulation of environmental conditions (Petchey et al. 2002; Gonzalez and Descamps-Julien 2004; Steiner et al. 2005) and only two of these involved the continuous variation in environmental conditions that make for stronger tests of the insurance hypothesis.

Continuous environmental fluctuations are desirable because, in competitive communities at least, it is environmental fluctuations that drives any stabilising effect of species richness (Ives et al. 1999; Gonzalez this volume). The negative covariances that stabilise community dynamics (see previous section) can result from interspecific competition and or differences in how species respond to environmental change (for example the growth rate of one species increases when temperature increases, while the growth rate of another species decreases). In theoretical communities of competitors, interspecific interactions increase negative covariance (stabilise) but at the same time increase species-level variance (destabilise). These contradictory effects of species richness on diversity cancel each other and the result is little net change in community level variability. The other source of negative covariance, difference in species' responses to changed environmental conditions (McNaughton 1977) will stabilise community dynamics if this increased richness introduces species that respond differently to environmental change (Ives et al. 1999). Ives et al. 1999 used a more mechanistic model of a competitive community where species interactions can alter population-level variances directly. Other more phenomenological approaches assume that species interactions can only indirectly affect population-level variances, via the effect of interactions on abundance and

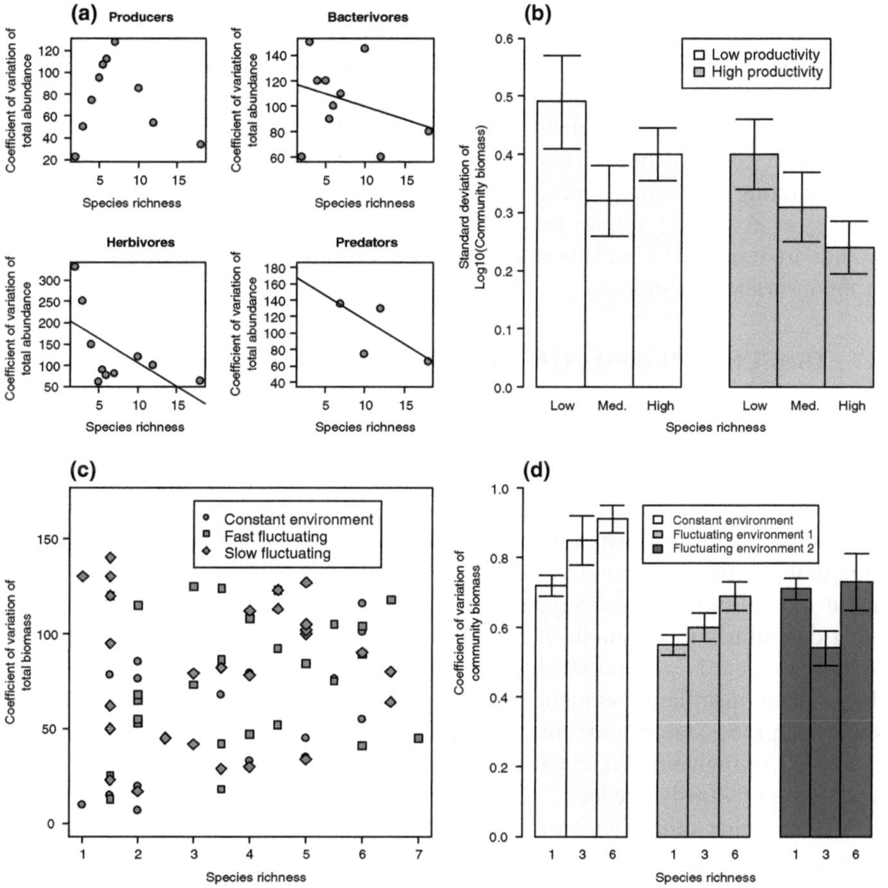

Figure 9.1. Relationships between species richness and temporal variability of total biomass of a group of species. (a) Temporal variability is stabilised by increased species richness in three of four functional groups (McGrady-Steed and Morin 2000). (b) Environmental productivity can influence stabilising effects of species richness (Steiner et al. 2005). (c) Environmental variability can have little effect on the relationship (Petchey et al. 2002), (d) or may in some cases reduce a destabilising effect of species richness on temporal variability (Gonzalez and Descamps-Julien 2004). Previously published graphs were digitised and reformatted with permission.

the relationship between a species' abundance and variance and cannot therefore reproduce the mechanistic results (Petchey et al. 2002).

In experiments with a constant environment treatment and a variable environment treatment, the prediction of the insurance effect is of a statistical interaction between species richness and environment, such that the stabilising effect of species richness is greater in variable environments than constant. This prediction arises because there is no environmental fluctuations for species to respond to differently in the constant environment, but there is in the fluctuating environment, where the insurance effect should operate. Gonzalez and Descamps-Julien (2004) reported a significant interaction between species richness and environmental fluctuations. In their experiment, species richness never had a stabilising effect on community variability, but the destabilising effect was arguably least in the variable environments (Figure 9.1d). Furthermore, summed covariances were significantly lower in fluctuating environments than constant ones. This provides some support for the insurance hypothesis operating through differences in species' responses to environmental change. Petchey et al. (2002) did not find a significant statistical interaction. Clearly, more studies are required before any statements about empirical evidence for mechanisms of the insurance hypothesis.

9.7 DISCUSSION AND FUTURE DIRECTIONS

There are surprisingly few studies of the insurance hypothesis of biodiversity that closely meet the assumptions made by the hypothesis and that are relatively straightforward to interpret. In particular this means studies that measure temporal variability (and not spatial) over multiple generations of the dominant organisms, and include a direct manipulation of species richness. I identified six such studies, and though there may be more, there are probably too few to make general conclusions about the validity of the insurance hypothesis, too few to make statements about what determines the strength of any insurance effect of biodiversity, and certainly too few to identify general mechanisms behind any insurance effect. Consequently, there seems insufficient evidence to conclude that species-rich communities often show increased stability at the community level, as has been hinted in previous reviews (Cottingham et al. 2001; Loreau et al. 2002; Worm and Duffy 2003). More studies are obviously needed.

It is true that other types of study fall outside the stringent criteria for inclusion as a case study in this article (e.g. Stachowicz et al. 2002) and that these are important for validating the insurance hypothesis and understanding relationships between diversity and stability. For example, observational studies can be important sources of information about whether the insurance hypothesis might stabilise in natural ecosystems and the importance of environmental variability (Johnson and Mann 1998; Romanuk and Kolasa 2002). Studies involving indirect manipulations of diversity (e.g. by fertiliser addition) can also be important for understanding the

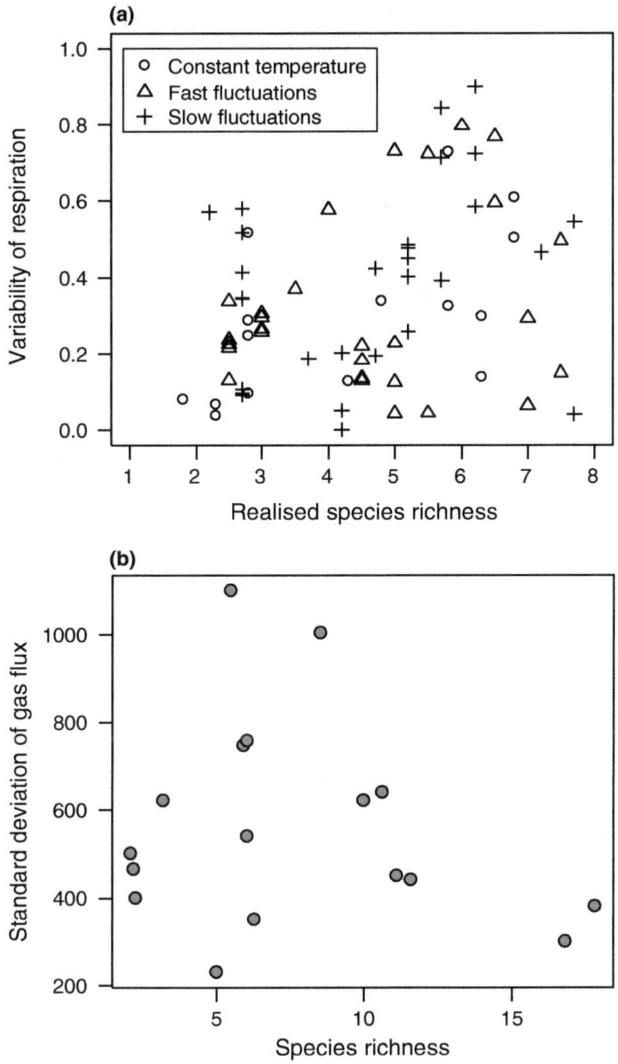

Figure 9.2. Relationships between species richness and temporal variability of an ecosystem process, here gas flux, which is a measure of community respiration. There is no evidence of the stabilising effect predicted by the insurance hypothesis in (a): previously unpublished data; or (b): Morin and McGrady-Steed (2004). Previously published graphs were digitised and reformatted with permission.

combined effect of environmental change and biodiversity change on ecosystem stability.

Apart from more long-term manipulative experiments with continuous environmental variability, what would enhance future research? A number of factors deserve exploration. First, evenness of species abundances is critical for altering any stabilising effects of diversity (Cottingham et al. 2001). Obviously, the variability of one dominant species will control community and ecosystem level variability, such that stabilising effects of additional rare species will be minimal. Experimental manipulations of evenness will be difficult, however, in long-term experiments because abundances can change greatly through time. Altering environmental conditions can influence evenness, but introduces a confounding variable. One approach is to report evenness patterns observed in experiments and attempt to use this information to better interpret results. For example, Steiner (2005) found that increasing species richness decreased evenness, so that perhaps the stabilising effect of species richness is weaker because of this.

An important design component of the case studies examined here is use of random combinations. The reasons for this are described above, but a consequence is that the experiments may provide relatively little information about effects of nonrandom loss of species. This is particularly important because real and likely extinctions are often of species with particular traits (e.g. larger body sizes). Furthermore, multiple studies show that nonrandom loss often has different and often more deleterious consequences for ecological systems than random loss (e.g. Nee and May 1997; Purvis et al. 2000; Petchey and Gaston 2002; Gross and Cardinale 2005). One way to simulate a particular order of species loss is to use dilution to decrease species richness (Giller et al. 2004). It is likely that more numerous species will be retained at high dilutions and low dilutions, whereas rare species may be absent from high dilution treatments. This implies that the richness gradient will contain a nested series of species compositions. In this kind of experiment, Romanuk et al. (2006) reported a stabilising effect of species richness that depended on nutrient enrichment. It might be interesting to see if the results of this experiment could be compared to results where the richness gradient is constructed by random selection of species.

Another route for future research is to manipulate components of biodiversity other than species richness. It is clear that species richness affects ecosystem properties through its effect on the distribution of species traits within a local community (Walker et al. 1999; Petchey 2004; Hooper et al. 2005). Experiments investigating effects of biodiversity on the magnitude of ecosystem properties have manipulated functional diversity as well as species richness (Hooper and Vitousek 1997; Hector et al. 1999) but there have not been such extensive examinations of the effects of functional diversity on temporal variability of ecosystem processes. High functional diversity should lead to lower temporal variability for at least two reasons. First, if high functional diversity implies greater differences in how species respond to environmental fluctuations than low functional diversity, one expects a strong stabilising effect due to Ives et al. (1999) mechanism. Second, temporal changes in composition should cause relatively little change in the magnitude of ecosystem functioning in high functional diversity communities because species are likely relatively redundant relative to each other. The ecosystem consequences of loss of one species are compensated for by a similar species. It is obvious therefore that the redundant species is only redundant on short timescales and becomes important for continued steady ecosystem functioning over longer timescales.

The functional consequences of biodiversity change need to be understood in a changing environment. Extinctions, the primary motivation for such research, now usually result from an environmental change, and therefore their consequence exists in this context. Most research removes an extinction from its cause and perhaps only studies half of the question about what changes in ecosystem functioning will occur in the future. Understanding the impact of species richness and environmental change on the temporal variability of community and ecosystem patterns and processes must be a priority for future research.

9.8 ACKNOWLEDGEMENTS

This chapter results from the 2005 Peter Yodzis Colloquium in Fundamental Ecology, "The influence of environmental noise on ecological systems". I am honoured to have participated and paid some tribute to Peter. The organisers, David Vasseur, Kevin McCann, and David Noakes, speakers, and delegates produced a unique, rich, and moving opportunity to discuss ecology. Thanks to Peter Morin, Jill McGrady-Steed, Andrew Gonzalez, and Chris Steiner for allowing me to recreate figures from their work. Comments by Dan Leary improved this work. It was partly funded by the Royal Society.

9.9 LITERATURE CITED

Allison, G.W. 1999. The implications of experimental design for biodiversity manipulations. *The American Naturalist* 153: 26–45.

Connell, J.H. and W.P. Sousa. 1983. On the evidence needed to judge ecological stability or persistence. *The American Naturalist* 121: 789–824.

Cottingham, K.L., B.L. Brown, and J.T. Lennon. 2001. Biodiversity may regulate the temporal variability of ecological systems. *Ecology Letters* 4: 72–85.

Daily, G. 1997. *Nature's services: societal dependence on natural ecosystems*. Island Press, Washington, DC.

Doak, D.F., D. Bigger, E.K. Harding, M.A. Marvier, R.E. O'Malley, and D. Thomson. 1998. The Statistical Inevitability of Stability-Diversity Relationships in Community Ecology. *The American Naturalist* 151: 264–276.

Elton, C.S. 1958. *The ecology of invasions by animals and plants*. Chapman & Hall, London.

Fukami, T., S. Naeem, and D.A. Wardle. 2001. On similarity among local communities in biodiversity experiments. *Oikos* 95: 340–348.

Gardner, M.R. and W.R. Ashby. 1970. Connectance of large dynamic (cybernetic) systems: critical values for stability. *Nature* 228: 784.

Gaston, K.J. and B.H. McArdle. 1994. The temporal variability of animal abundances: measures, methods and patterns. *Philosophical Transactions of the Royal Society of London, Series B, Biological Sciences* 345: 335–358.

Giller, P.S., H. Hillebrand, U.G. Berninger, M.O. Gessner, S. Hawkins, P. Inchausti, C. Inglis, H. Leslie, B. Malmqvist, M.T. Monaghan, P.J. Morin, and G. O'Mullan. 2004. Biodiversity effects on ecosystem functioning: emerging issues and their experimental test in aquatic environments. *Oikos* 104: 423–436.

Gonzalez, A. and B. Descamps-Julien. 2004. Population and community variability in randomly fluctuating environments. *Oikos* 106: 105–116.

Gross, K. and B.J. Cardinale. 2005. The functional consequenes of random vs. ordered species extinctions. *Ecology Letters* 8: 409–418.

Hector, A., B. Schmid, C. Beierkuhnlein, M.C. Caldeira, M. Diemer, P.G. Dimitrakopoulos, J.A. Finn, H. Freitas, P.S. Giller, J. Good, R. Harris, P. Högberg, K. Huss-Danell, J. Joshi, A. Jumpponen, C. Körner, P.W. Leadley, M. Loreau, A. Minns, C.P.H. Mulder, G. O'Donovan, S.J. Otway, J.S. Pereira, A. Prinz, D.J. Read, M. Scherer-Lorenzen, E.-D. Schulze, A.-S.D. Siamantziouras, E.M. Spehn, A.C. Terry, A.Y. Troumbis, F.I. Woodward, S. Yachi, and J.H. Lawton. 1999. Plant diversity and productivity experiments in European grassland. *Science* 286: 1123–1127.

Hooper, D.U., F.S. Chapin, J.J. Ewel, A. Hector, P. Inchausti, S. Lavorel, J.H. Lawton, D.M. Lodge, M. Loreau, S. Naeem, B. Schmid, H. Setälä, A.J. Symstad, J. Vandermeer, and D.A. Wardle. 2005. Effects of biodiversity on ecosystem functioning: a consensus of current knowledge. *Ecological Monographs* 75: 3–35.

Hooper, D.U. and P.M. Vitousek. 1997. The effects of plant composition and diversity on ecosystem processes. *Science* 277: 1302–1305.

Hughes, J.B. and J. Roughgarden. 1998. Aggregate community properties and the strength of species' interactions. *Proceedings of the National Academy of Sciences of the USA* 95: 6837–6842.

Hughes, J.B. and J. Roughgarden. 2000. Species diversity and biomass stability. *The American Naturalist* 155: 618–627.

Huston, M.A. 1997. Hidden treatments in ecological experiments: re-evaluating the ecosystem function of biodiversity. *Oecologia* 110: 449–460.

Ives, A.R. 1995. Predicting the response of populations to environmental change. *Ecology* 76: 926–941.

Ives, A.R., K. Gross, and J.L. Klug. 1999. Stability and variability in competitive ecosystems. *Science* 286: 542–544.

Jessup, C.M., R. Kassen, S.E. Forde, B. Kerr, A. Buckling, P.B. Rainey, and B.J.M. Bohannan. 2004. Big questions, small worlds: microbial model systems in ecology. *Trends in Ecology & Evolution* 19: 189–197.

Johnson, C.R. and K.H. Mann. 1988. Diversity, patterns of adaptation, and stability of Nova Scotian Kelp beds. *Ecological Monographs* 58: 129–154.

Lehman, C.L. and D. Tilman. 2000. Biodiversity, stability, and productivity in competitive communities. *The American Naturalist* 156: 534–552.

Loreau, M., A.L. Downing, M.C. Emmerson, A. Gonzalez, J. Hughes, P. Inchausti, J. Joshi, J. Norberg, and O. Sala. 2002. A new look at the relationship between diversity and stability. In: *Biodiversity and ecosystem functioning: synthesis and perspectives* (eds., M. Loreau, S. Naeem, and P. Inchausti), pp. 79–91. Oxford University Press, Oxford.

Loreau, M., S. Naeem, P. Inchausti, J. Bengtsson, J.P. Grime, A. Hector, D.U. Hooper, M.A. Huston, D. Raffaelli, B. Schmid, D. Tilman, and D.A. Wardle. 2001. Biodiversity and ecosystem functioning: current knowledge and future challenges. *Science* 294: 804–808.

MacArthur, R. 1955. Fluctuations of animal populations, and a measure of community stability. *Ecology* 36: 533–536.

May, R.M. 1972. Will a large complex system be stable? *Nature* 238: 413–414.

McCann, K.S. 2000. The diversity-stability debate. *Nature* 405: 228–233.

McGrady-Steed, J., P.M. Harris, and P.J. Morin. 1997. Biodiversity regulates ecosystem predictability. *Nature* 390: 162–165.

McGrady-Steed, J. and P.J. Morin. 2000. Biodiversity, density compensation, and the dynamics of populations and functional groups. *Ecology* 81: 361–373.

McNaughton, S.J. 1977. Diversity and stability of ecological communities: a comment on the role of empiricism in ecology. *The American Naturalist* 111: 515–525.

Morin, P.J. 1998. Realism, precision, and generality in experimental ecology. In: *Experimental ecology: issues and perspectives* (eds., W.J.J. Resetarits and J. Bernardo), pp. 236–253. Oxford University Press, Oxford.

Morin, P.J. and J. McGrady-Steed. 2004. Biodiversity and ecosystem functioning in aquatic microbial systems: a new analysis of temporal variation and species richness-predictability relations. *Oikos* 104: 458–466.

Naeem, S. 1998. Species redundancy and ecosystem reliability. *Conservation Biology* 12: 39–45.

Naeem, S. 2002. Disentangling the impacts of diversity on ecosystem functioning in combinatorial experiments. *Ecology* 83: 2925–2935.

Naeem, S. and S. Li. 1997. Biodiversity enhances ecosystem reliability. *Nature* 390: 507–509.

Nee, S. and R.M. May. 1997. Extinction and the loss of evolutionary history. *Science* 278: 692–694.

Petchey, O.L. 2000. Prey diversity, prey composition, and predator population stability in experimental microcosms. *Journal of Animal Ecology* 69: 874–882.

Petchey, O.L. 2004. On the statistical significance of functional diversity. *Functional Ecology* 18: 297–303.

Petchey, O.L., T.J. Casey, L. Jiang, P.T. McPhearson, and J. Price. 2002. Species richness, environmental fluctuations, and temporal change in total community biomass. *Oikos* 99: 231–240.

Petchey, O.L. and K.J. Gaston. 2002. Extinction and the loss of functional diversity. *Proceedings of the Royal Society of London, Series B, Biological Sciences* 269: 1721–1727.

Pfisterer, A.B. and B. Schmid. 2002. Diversity-dependent production can decrease the stability of ecosystem functioning. *Nature* 416: 84–86.

Pimm, S.L. 1984. The complexity and stability of ecosystems. *Nature* 307: 321–326.

Purvis, A., P.-M. Agapow, J.L. Gittleman, and G.M. Mace. 2000. Nonrandom extinction and the loss of evolutionary history. *Science* 288: 328–330.

Romanuk, T.N. and J. Kolasa. 2002. Environmental variability alters the relationship between richness and variability of community abundances in aquatic rock pool microcosms. *Ecoscience* 9: 55–62.

Romanuk, T.N., R.J. Vogt, and J. Kolasa. (2006) Nutrient enrichment weakens the stabilizing effect of species richness. *Oikos* 114: 291–302.

Roscher, C., J. Schumacher, J. Baade, W. Wilcke, G. Gleixner, W.W. Weisser, B. Schmid, and E.D. Schulze. 2004. The role of biodiversity for element cycling and trophic interactions: and experimental approach in a grassland community. *Basic and Applied Ecology* 5: 107–121.

Roscher, C., V.M. Temperton, M. Scherer-Lorenzen, M. Schmitz, J. Schumacher, B. Schmid, N. Buchmann, W.W. Weisser, and E.D. Schulze. 2005. Overyielding in experimental grassland communities – irrespective of species pool or spatial scale. *Ecology Letters* 8: 419–429.

Schmid, B., A. Hector, M.A. Huston, P. Inchausti, I. Nijs, P.W. Leadley, and D. Tilman. 2002. The design and analysis of biodiversity experiments. In: *Biodiversity and ecosystem functioning: synthesis and perspectives* (eds., M. Loreau, S. Naeem, and P. Inchausti), pp. 61–75. Oxford University Press, Oxford.

Schwartz, M.W., C.A. Brigham, J.D. Hoeksema, K.G. Lyons, M.H. Mills, and P.J. van Mantgem. 2000. Linking biodiversity to ecosystem function: implications for conservation ecology. *Oecologia* 122: 297–305.

Srivastava, D.S., J. Kolasa, J. Bengtsson, A. Gonzalez, S.P. Lawler, A.R. Miller, P. Munguia, T. Romanuk, D.C. Schneider, and M.K. Trzcinski. 2004. Are natural microcosms useful model systems for ecology. *Trends in Ecology & Evolution* 19: 379–384.

Stachowicz, J.J., H. Fried, R.W. Osman, and R.B. Whitlatch. 2002. Biodiversity, invasion resistance, and marine ecosystem function: reconciling patter and process. *Ecology* 83: 2575–2590.

Steiner, C.F., Z.T. Long, J.A. Krumins, and P.J. Morin. 2005. Temporal stability of aquatic food webs: partitioning the effects of species diversity, species composition and enrichment. *Ecology Letters* 8: 819–828.

Tilman, D. 1996. Biodiversity: population versus ecosystem stability. *Ecology* 77: 350–363.

Tilman, D., C.L. Lehman, and C.E. Bristow. 1998. Diversity–stability relationships: statistical inevitability or ecological consequence? *The American Naturalist* 151: 277–282.

Tilman, D., P.B. Reich, J. Knops, D. Wedin, T. Mielke, and C.L. Lehman. 2001. Diversity and productivity in a long-term grassland experiment. *Science* 294: 843–845.

Walker, B., A. Kinzig, and J. Langridge. 1999. Plant attribute diversity, resilience, and ecosystem function: the nature and significance of dominant and minor species. *Ecosystems* 2: 95–113.
Wardle, D.A. 1998. A more reliable design for biodiversity study. *Nature* 394: 30.
Wardle, D.A. 2002. *Communities and ecosystems: linking the aboveground and belowground components.* Princeton University Press, Princeton.
Worm, B. and J.E. Duffy. 2003. Biodiversity, productivity and stability in real food webs. *Trends in Ecology & Evolution* 18: 628–632.
Yachi, S. and M. Loreau. 1999. Biodiversity and ecosystem productivity in a fluctuating environment: the insurance hypothesis. *Proceedings of the National Academy of Sciences of the USA* 96: 1463–1468.

CHAPTER 10

ENVIRONMENTAL VARIABILITY AND THE ANTARCTIC MARINE ECOSYSTEM

VALERIE LOEB

Moss Landing Marine Laboratories, 8272 Moss Landing Road, Moss Landing, California, USA, 95039
Phone: (831) 771-4476, Fax: (831) 632-4403, E-mail: loeb@mlml.calstate.edu

David A. Vasseur and Kevin S. McCann, The Impact of Environmental Variability on Ecological Systems, 197–225. © 2007 *Springer.*

10.1 ABSTRACT

This chapter presents an overview of the physical environment and pelagic marine ecosystem of the Southern Ocean with special attention to the Antarctic Peninsula region, south of South America. The Antarctic Peninsula region has been well studied because it is an important feeding, spawning, and nursery ground for Antarctic krill, *Euphausia superba*, the keystone species for the Antarctic marine food web. The physical and biological data sets collected in this region allow examination of seasonal, interannual and longer-scale ecosystem variability and the coupled atmospheric-oceanic-ice processes that underlie this variability. Evidence is presented that supports the hypothesis that this region is strongly impacted by meridional atmosphere teleconnections instigated in the western tropical Pacific Ocean by El Niño–Southern Oscillation (ENSO) variability that exhibits a 3- to 5-year periodicity.

Keywords: El Niño–Southern Oscillation, sea ice, krill, salps, Antarctic ecosystem.

10.2 INTRODUCTION

Two factors promoting evolution of Antarctic marine ecosystems – currents and ice – remain primary forces underlying ecosystem structure, function, and variability today. Wind-driven current systems and sea ice determine three circumpolar faunal zones: the northern ice free zone dominated by copepods and salp *Salpa thompsoni*; seasonal sea ice zone dominated by Antarctic krill, *Euphausia superba*; and southern permanent pack ice zone dominated by "ice krill" *E. crystallorophias* (Hempel 1985). Evolution of endemic species comprising communities in these zones started about 25 million years ago (mya) following the opening of Drake Passage when circulation of the Antarctic Circumpolar Current (ACC) permitted enhanced glaciation and the Polar Front (Antarctic Convergence) became a barrier between cold waters and subtropical water to the north (Eastman 1993; Clarke 1990; Clarke et al. 2005). Biological, physiological, and ecological adaptations by species to seasonal, annual, decadal, and larger-scale variations in currents and sea ice over millions of years undoubtedly have resulted in co-evolved communities and ecosystems within each of the faunal zones. The most studied of these is the "krill-centric" ecosystem of the seasonal sea ice zone characterized by a food web largely dependent on the filter feeding *E. superba* as a link between phytoplankton and higher trophic levels.

Recent studies have indicated significant decreases in krill population size and increased salp concentrations, particularly at higher latitudes, that have been linked to decreasing sea ice extent and atmospheric warming (Atkinson et al. 2004). Additionally, seasonal and interannual fluctuations in concentrations of copepods and salps in the Antarctic Peninsula region indicate the importance of atmospherically driven circulation processes underlying population dynamics of these biomass dominant taxa and krill recruitment success (Loeb et al., submitted).

This work focuses on krill, salp, and copepod dynamics as reflections of a long evolutionary history within the Southern Ocean ecosystem. It then describes seasonal, interannual, and longer-term abundance variations of these taxa in relation to hydrography and sea ice dynamics primarily in the Antarctic Peninsula and South Atlantic sector. These in turn are related to larger scale coupled atmospheric-sea ice-ocean dynamics teleconnections instigated in the western tropical Pacific Ocean by El Niño–Southern Oscillation (ENSO) variability (McPhaden, Chapter 1, this volume) that ultimately drive ecosystem variability here and that also may explain aspects of the evolved life histories of these dominant taxa.

10.3 EVOLUTIONARY HISTORY OF THE SOUTHERN OCEAN

Isolation of the Antarctic continent occurred about 25 mya when Drake Passage opened allowing the Antarctic Circumpolar Current to become fully developed (Lawver and Gahagan 2003; Clarke et al. 2005). This then set up conditions that resulted in the biogeographical isolation of the Southern Ocean and separation of *E. superba* and *E. crystallorophias* around 3.1–2.8 mya (Patarnello et al. 1996; Zane and Patarnello 2000), suggesting formation of their respective seasonal sea ice and permanent pack ice habitats during this time. Periods of warming and cooling and attendant glacial retreat and advance have regularly occurred on about 100,000-year cycles over at least the past million years (Petit et al. 1999). Major changes in krill distribution and population size are believed to have been associated with these cycles (Spiridonov 1996; Zane et al. 1998) implying that long-term and large-scale variations in sea ice extent, atmospheric and oceanic circulation processes underlie the evolutionary history and current structure of Antarctic marine ecosystems.

10.4 GENERAL HYDROGRAPHY OF THE SOUTHERN OCEAN

The Southern Ocean is composed of the eastward flowing Antarctic Circumpolar Current (ACC; the "West Wind Drift") and narrow westward flowing Antarctic Coastal Current ("East Wind Drift"). Much of the transport of the ACC is carried by a series of relatively narrow frontal jets the most notable of which are the Subantarctic, Polar, and Southern Antarctic Circumpolar Current Fronts (Orsi et al. 1995; Pollard et al. 2002; Figures 10.1 and 10.2). The Polar Front (formerly Antarctic Convergence) acts as the approximate northern boundary for the Southern Ocean. Its development was important in isolating the evolving fauna and few present epipelagic species transgress it. In this respect the Polar Front is considered to delimit the Antarctic biogeographic region (Clarke et al. 2005). The southernmost front associated with the ACC is the southern ACC front (sACCf) and the southernmost limit of ACC-derived waters is defined by the southern ACC boundary (Bndy; Orsi et al. 1995).

At depths greater than 500 m nutrient-rich, warm, and saline Circumpolar Deep Water (CDW) occurs (Sievers and Nowlin 1984). This water mass shoals to about 200 m at the southern portion of the ACC where the sACCf and Bndy are found in

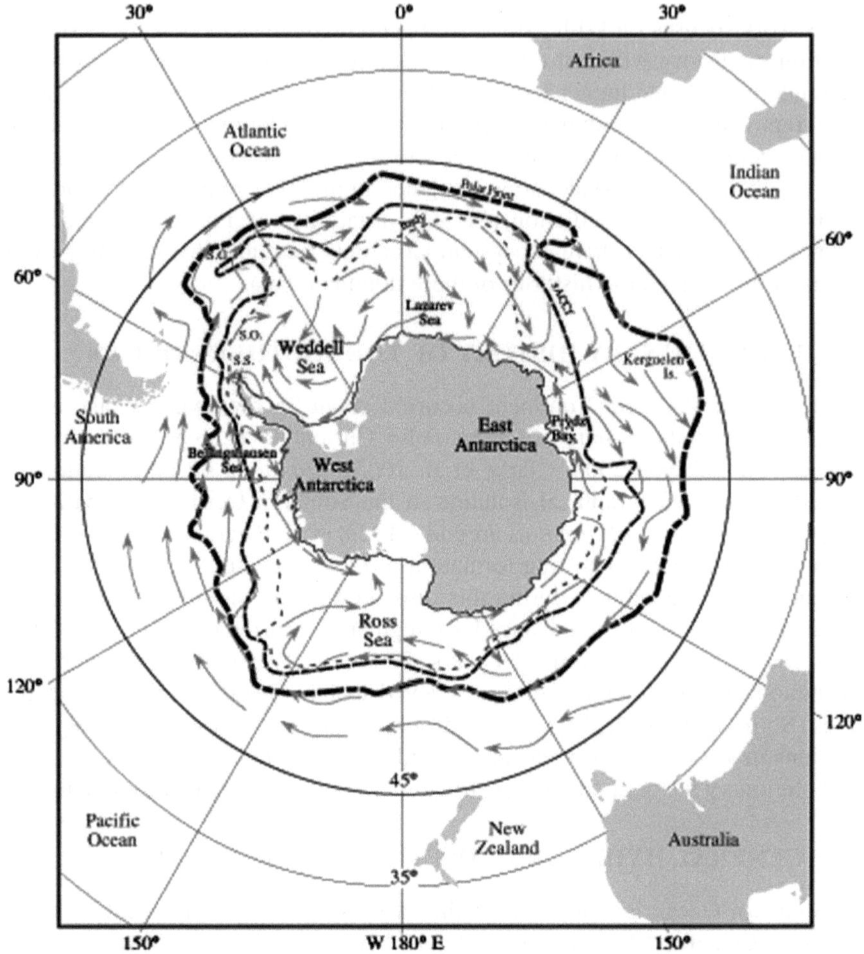

Figure 10.1. Surface circulation of the Southern Ocean highlighting major features of the Antarctic Circumpolar Current (ACC) and coastal gyres. Heavy and light dashed lines poleward of the Polar Front indicate the southern ACC Front (sACCf) and Southern Boundary of the ACC (Bndy). Island names: S.G. South Georgia; S.O. South Orkneys; S.S. South Shetlands.

proximity to the shelf/slope region, such as occurs in the Antarctic Peninsula region. Enhanced primary and secondary productivity have been associated with the sACCf and Bndy in the regions where CDW shoals (Tynan 1998; Ward et al. 2002, 2003; Prézelin et al. 2004).

South of the ACC are regionally-based gyres, the largest of which are in the Weddell and Ross Seas (Figure 10.1). These gyres are effective conduits for transport of heat, salt (Dinniman et al. 2003; Fahrbach et al. 2004), and nutrients (Smith et al. 2003) to the continental margins. Smaller clockwise gyres are located in the

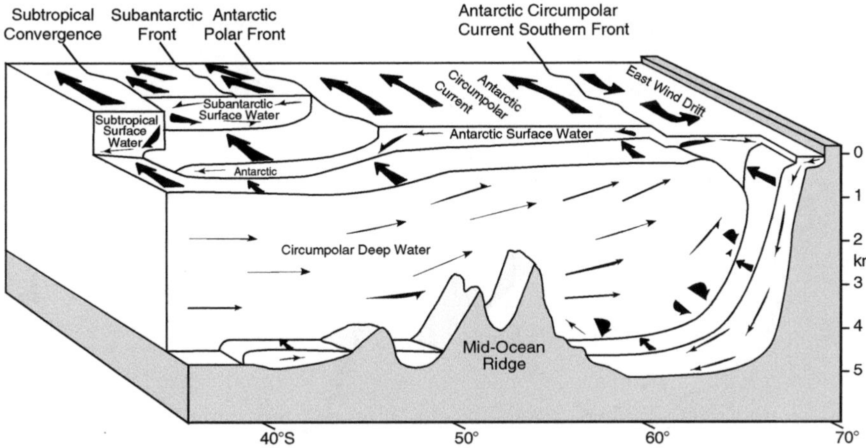

Figure 10.2. Generalized hydrographic features of the Southern Ocean. Large arrows indicate lateral movement of water types; thin arrows are inferred vertical movement. (Modified from Lutjeharms et al. 1985).

Lazarev Sea, Kerguelen Plateau, Prydz Bay and West Antarctic Peninsula (Amos 1984). Localized gyres also occur over the Antarctic continental shelf (Stein 1992; Ichii et al. 1998; Smith et al. 1999). These gyres provide recirculation systems and as such produce biological retention zones (Spiridonov 1996). The Antarctic continental shelf region and southern part of the ACC are subject to seasonal cycles of solar insolation and temperature that drive ice formation and retreat and comprise the seasonal sea ice zone.

10.5 SEA ICE

Sea ice growth and decay are affected by atmospheric and oceanic factors including surface air temperature, wind, ocean currents, and sea surface temperature (Zwally et al. 2002). Minimum and maximum ice cover occurs, respectively, in February–March and August–October and involves melting and freezing about 75–80% of the sea ice each year. The seasonal sea ice zone, defined by the northern and southern ice extent in winter/spring and summer/fall, includes the northern branch of the Weddell gyre and the Antarctic Peninsula and is therefore broadest and extends furthest north in the Atlantic sector (Figure 10.3; Hempel 1985). For average conditions, winter sea ice covers more than half of the open water area south of the Polar Front (El Sayed 1985). However, large interannual and longer-term variability in sea ice concentration and extent result from large-scale atmospheric and ocean circulation effects (Zwally et al. 2002; Comiso 2003).

The variability in sea ice concentration and extent observed for the Southern Ocean impose strong seasonality on the upper water column. This affects spring plankton blooms and production cycles that are dependent on the timing of sea ice

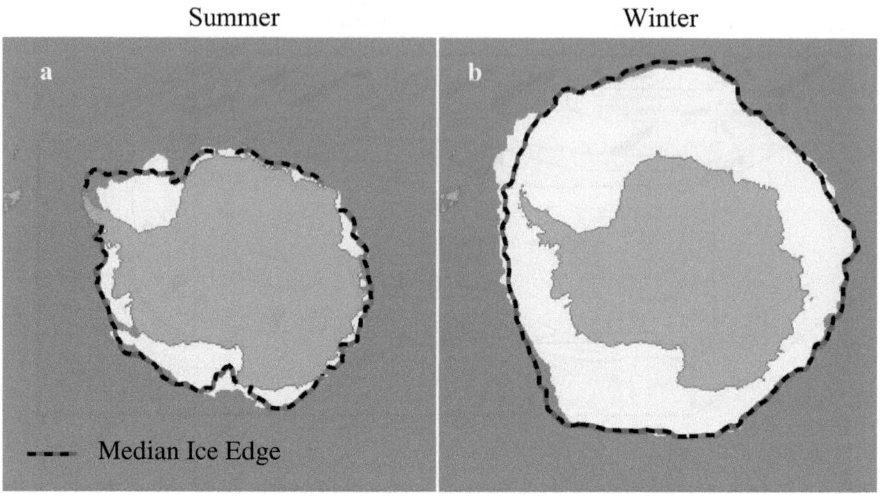

Figure 10.3. Seasonal sea ice extent during summer (March 1993) and winter (September 1995) relative to the long-term median. The months and years presented here show extremes in below and above average sea ice extent in the Antarctic Peninsula region during the respective seasons.

retreat with subsequent water column stabilization and seeding by ice algae (Eicken 1992; Spiridonov 1995). The underside of sea ice also provides food supplies in the form of concentrated shade-adapted phytoplankton (ice algae) and heterotrophs as well as refuge from predators and is an essential component of the krill habitat (Smetacek et al. 1990; Eicken 1992; Schnack-Schiel 2003). Sea ice also constitutes a critical habitat for top predators including seals, penguins, other seabirds and whales (Ainley et al. 2003).

10.6 DOMINANT ZOOPLANKTON COMPONENTS OF SOUTHERN OCEAN ECOSYSTEMS

The Antarctic circumpolar pelagic habitats described by Spiridonov (1996) and faunal zones describe by Hempel (1985) generally correspond and consist of an ice free ACC that supports a diverse zooplankton assemblage dominated by herbivorous copepods and the salp *Salpa thompsoni* (Figure 10.4a,b), a seasonal sea ice zone that supports greatest concentrations of *Euphausia superba* (Figure 10.4c), and a pack ice zone dominated by "ice krill" *E. crystallorophias*. The vast majority of information on Southern Ocean pelagic marine ecosystems is derived from the ice free and seasonal sea ice zones dominated by copepods, salps and krill.

Copepods, salps and krill are able to maintain their general positions within their respective zones through seasonal, diel, and ontogenetic vertical migrations that expose them to differential transport with depth (Figure 10.2; Longhurst 1976; Laws 1985; Knox 1994). During summer the biomass dominant copepod species

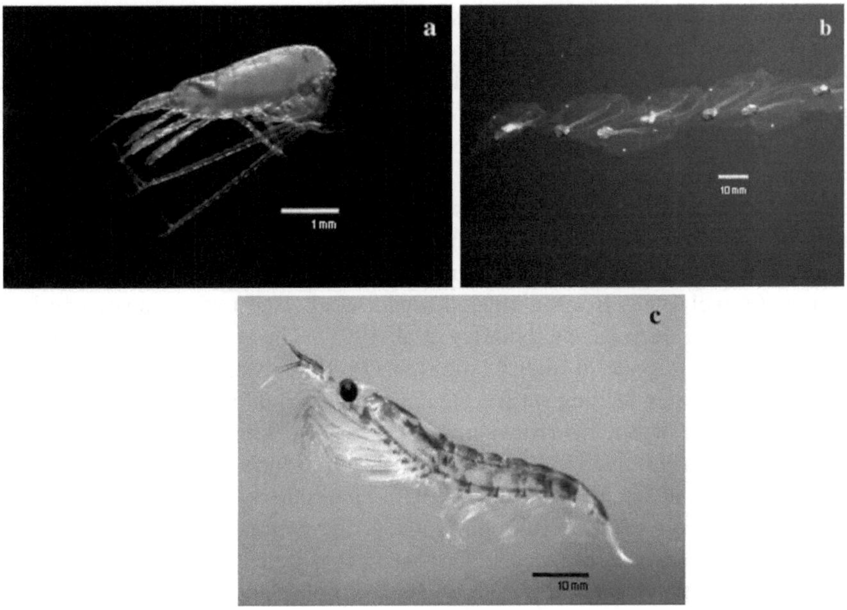

Figure 10.4. Important Antarctic zooplankton taxa: (a) large oceanic copepod *Calanus propinquus* (V. Loeb); (b) a chain of *Salpa thompsoni* aggregates (Dirk Schories); (c) adult Antarctic krill, *Euphausia superba* (Rob Rowley).

concentrate in ACC surface waters where they are subjected to northward Ekman drift (Hempel 1985) but in winter they descend to 500–1,000 m within the southward moving CDW. Spawning is associated with the spring ascent supplying offspring to surface layers during summer.

Salpa thompsoni undergoes seasonal alternation of generations with the asexual solitary stage overwintering throughout the water column but with maximum concentrations probably at depths >500 m (Foxton 1966; Lancraft et al. 1991) which could allow southward transport in association with CDW. These grow, mature, and rise to surface layers during spring and summer months where they release chains of the sexual aggregate stage. Aggregate individuals grow and produce the solitary stage primarily in autumn. During spring and summer both stages exhibit diel vertical migrations from depths ≥600 m in day to surface layers at night, constantly filter-feeding on small particles as they move (Foxton 1966; Casereto and Nemoto 1986; Lancraft et al. 1989). Given optimal feeding and water column conditions massive concentrations of aggregates can develop during spring and summer months (Foxton 1966; Daponte et al. 2001). High phytoplankton concentrations and ice cover appear to be unfavorable for *S. thompsoni* (Nicol et al. 2000).

Unlike copepods and salps which live less than a year, krill can live up to five and possibly as long as 8 years (Siegel 1987). Reproductively mature krill migrate offshore to shelf break and slope areas where they mate and spawn eggs over deep waters. This strategy increases the probability of the dense embryo encountering warm CDW which favors accelerated development, prevents the embryo from reaching the sea floor prior to hatching and reduces the depth range for larval ascent (Fraser 1936; Marr 1962; Siegel 1988; Hofmann et al. 1992; Lascara et al. 1999; Hofmann and Hüsrevoglu 2003). Larvae reach surface waters as first feeding early calyptopis stages 15–30 days after spawning, dependent on ambient temperature conditions (Hofmann et al. 1992). During spring and summer 1-year-old juveniles remain furthest onshore while immature and small mature stages inhabit intermediate areas (Siegel 1988; Brinton 1991; Lascara et al. 1999) where spawning by the latter occurs over deep basins (Huntley et al. 1991). These stages followed by post-spawning adults migrate to higher latitudes with the advancing season and remain there through winter feeding on ice algae and/or available heterotrophic resources including benthic organisms (Kawaguchi et al. 1986; Hofmann and Lascara 2000). Late larval stages entrained in continental shelf gyres during fall are likely to inhabit areas with seasonal sea ice cover providing a favorable overwintering habitat (food and sanctuary) and good recruitment success the following spring. Survivorship of individuals entrained in the southern ACC is dependent on sufficient supplies of pelagic phytoplankton, sea ice algae, heterotrophic and detrital particles and is enhanced by transport to areas covered by sea ice in fall and winter (Fach et al. 2002; Hofmann and Murphy 2004).

The krill reproductive range includes the sACCf, Weddell and Ross Sea Gyres, and mesoscale eddy systems in the Bellingshausen Sea, Prydz Bay and D'Urville Sea (Spiridonov 1996). Areas north of the sACCf are sources of expatriation while those to the south are zones of recirculation and retention (Figure 10.5). On a circumpolar scale krill density both north and south of the sACCf is related to food supply which in turn depends on sea ice, circulation and nutrients (particularly iron) that promote primary production near shelves, undersea ridges and other bathymetric features, at fronts and along the ice edge (Holm-Hansen et al. 1994, 1997, 2004; de Baar et al. 1996; Pollard et al. 2002; Atkinson et al. 2004). Since these factors vary regionally krill concentrations are irregularly distributed around the continent (Figure 10.5) with over 50% of the Southern Ocean stock represented within the southwest Atlantic sector (i.e., West Antarctic Peninsula to east of the Weddell Sea). Krill advected away from this sector by the sACCf are believed to be the source of regional stocks associated with retention zones and ecosystems downstream (Spiridonov 1996).

10.7 ECOLOGICAL IMPORTANCE OF THE SOUTHWEST ATLANTIC SECTOR

Within the Southwest Atlantic sector the South Shetland-Elephant Island region and adjacent Bransfield Strait and Drake Passage waters (Figure 10.5) form an important krill swarming, spawning and nursery area (Marr 1962; Spiridonov

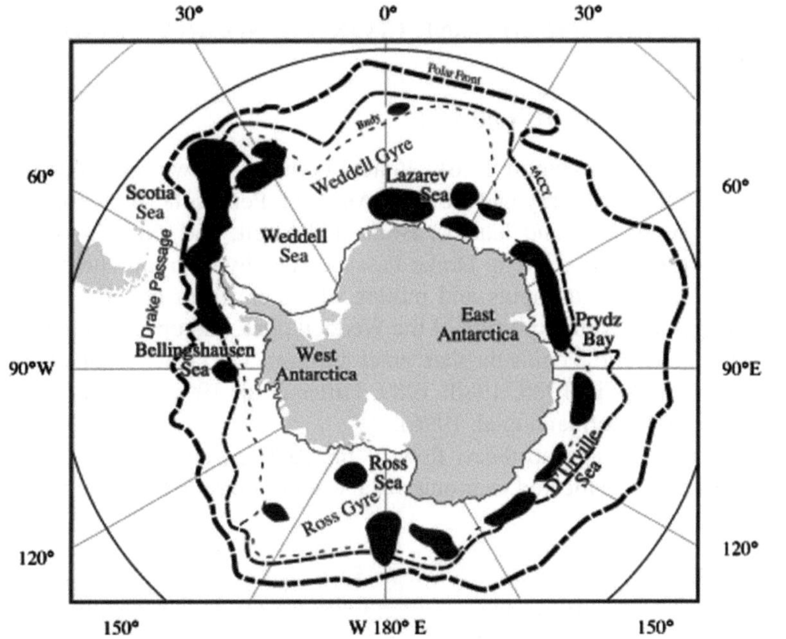

Figure 10.5. Circum-Antarctic oceanographic features and Antarctic krill distribution showing the important advection-retention zones and areas of elevated krill concentrations. (Modified from Spiridonov 1996.) sACCf is the Southern Front of the ACC and Bndy is the Southern Boundary of the ACC.

1996). Here are mixtures of individuals advected from the Bellingshausen and Weddell Seas and from local production within Bransfield and Gerlache Straits (Brinton 1991; Huntley and Brinton 1991; Spiridonov 1996). This area and the Weddell Gyre are the major sources of krill to land-based predator populations at the South Orkney, South Sandwich, and South Georgia Islands and probably replenish stocks in retention zones around the Antarctic continent (Spiridonov 1996). Because of this importance the South Shetland-Elephant Island area has a long history of scientific research that provides data essential for detecting and evaluating ecosystem variability over a range of time scales. These include the 1925–1939 *Discovery* Expeditions, Biological Investigations of Marine Antarctic Systems and Stocks (BIOMASS) Experiment surveys of 1980–1981 and 1983–1985 and annual surveys of the US Antarctic Marine Living Resources (AMLR) Program from 1990 to present. Additional information from research focused in the West Antarctic Peninsula by the Palmer Station Long-Term Ecological Research (LTER) Program (Ross et al. 1996; Quetin and Ross 2003) and South Georgia by British Antarctic Survey together with modeling studies over the past decade support ecosystem coherency across the South Atlantic sector (Reid and Croxall 2001; Reid et al. 2002; Ward et al. 2002; Fraser and Hofmann 2003; Atkinson et al. 2004; Hofmann and Murphy 2004; Murphy et al. 2004a,b).

10.8 THE ANTARCTIC MARINE LIVING RESOURCES (AMLR) PROGRAM

10.8.1 Study area and hydrography

The AMLR study area (Figure 10.6), situated between the western Weddell-Scotia Confluence (WSC) boundary and the West Antarctic Peninsula LTER survey region, includes the South Shetland Islands, Bransfield Strait, and ridge associated with the Shackleton Fracture Zone in Drake Passage. The hydrography here is complex and variable and reflects inputs and mixing of waters from the ACC in southern Drake Passage, the western portion of the Weddell gyre, and upstream regions along the western Antarctic Peninsula that enter through Gerlache Strait and western Bransfield Strait (Stein 1986, 1988, 1989; Niiler et al. 1991; Capella et al. 1992; Garcia et al. 1994; Hofmann et al. 1996).

Within Drake Passage northeast flow of the ACC parallels the South Shetland Islands with the Bndy located near outer island shelf areas up to the point northwest

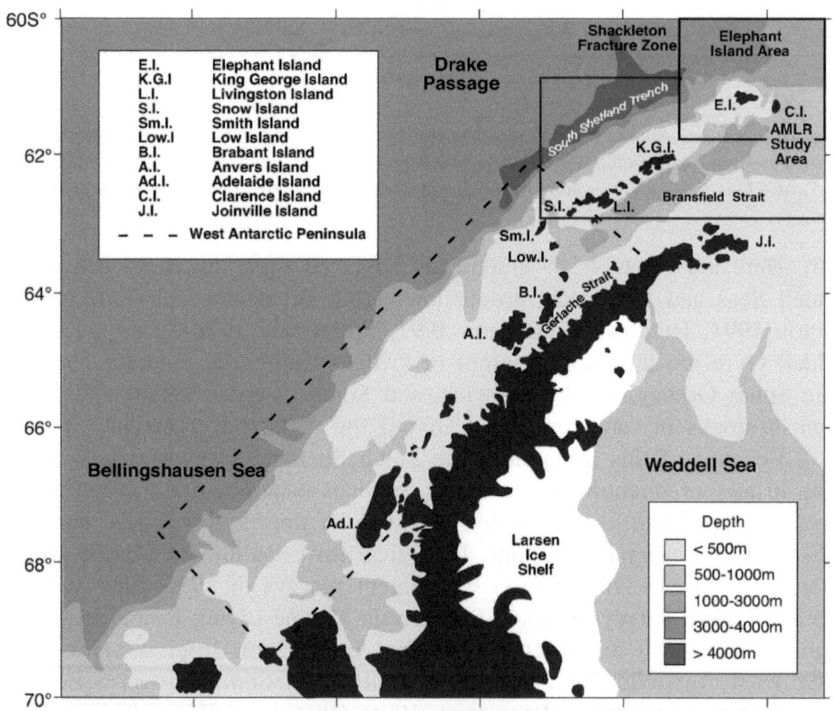

Figure 10.6. Map of the Antarctic Peninsula region indicating place names, important bathymetric features, the Antarctic Marine Living Resources (AMLR) Program survey area and Elephant Island area used for long-term time series analyses.

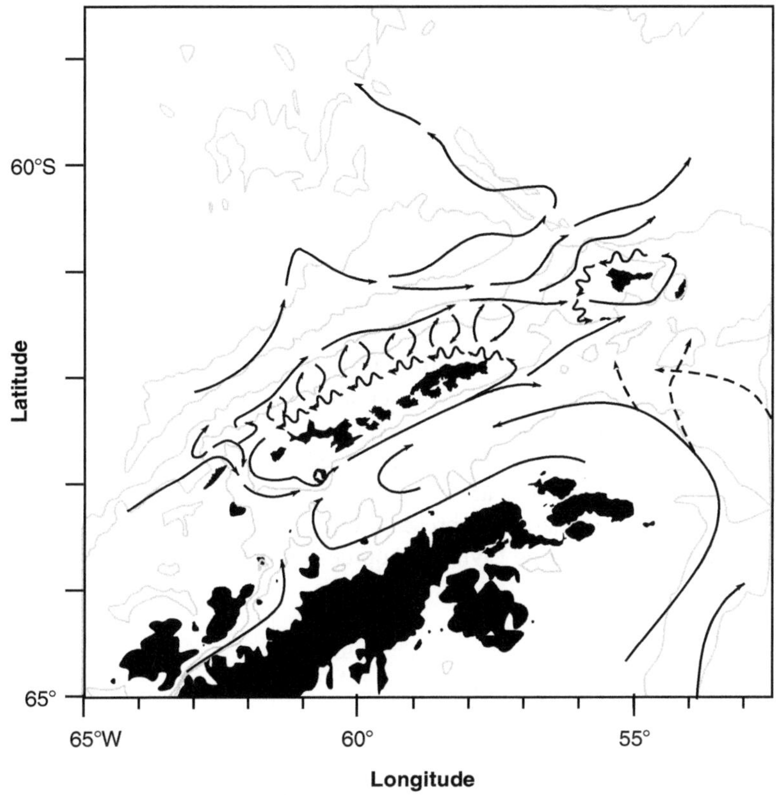

Figure 10.7. Generalized surface circulation patterns in the AMLR survey area, showing deflection of the ACC due to bathymetry of the Shackleton Fracture Zone, variable flow into Bransfield Strait from the Weddell Gyre (dashed lines) and gyres over island shelves and basins within Bransfield Strait.

of Elephant Island where the ACC is deflected northward by the Shackleton Fracture Zone ridge (Figure 10.7; Amos 2001). The island shelf areas are influenced by counterclockwise flows around Livingston/King George Islands and Elephant Island (Figure 10.6). Retention within these areas is further enhanced by small scale eddies along the north shelf break (Ichii et al. 1998) and exchange between north and south shelf waters through straits separating the islands (Clowes 1934).

10.9 METHODS

The AMLR Program has conducted multidisciplinary ship-based sampling operations during generally two austral summer (January–March) surveys in the South Shetlands Island area since 1990. Although the survey design has changed somewhat over the years the Elephant Island area (Figure 10.7) has been consistently occupied and represents a historically sampled area used for long-term analyses of the Antarctic Peninsula marine ecosystem (Siegel and Loeb 1995; Loeb et al. 1997). Sampling is conducted at about 100 fixed stations located about 40 km apart on north-south transects during each cruise leg (Figure 10.7). Hydrographic and primary production data are collected by CTD/Carousel casts made to 750 m or about 5 m off the bottom in shallower water. Hydrographic zones are characterized according to temperature and salinity relationships of waters representing the ACC, Weddell Sea, and three intermediate mixtures (Amos 2001). Krill and zooplankton are collected by 170 m open oblique hauls using a 1.8 m Isaacs-Kidd Midwater Trawl fitted with 505 μm mesh plankton net, flow meter, and real-time depth monitor. Fresh or freshly frozen samples are processed onboard within several hours of collection. Entire samples or representative subsamples of at least 100 krill and 100 salps are measured and maturity stages identified according to Makarov and Denys (1981) and Foxton (1966), respectively. Subsamples of all remaining taxa are identified to species if possible and enumerated. Abundance is expressed as numbers per 1,000 m^{-3} water filtered. Integrated 0–100 m Chl-a data are used to assess primary production for each cruise. Sampling specifics are presented in Siegel and Loeb (1995). These and the surveys reported here are also presented in a series of annual reports available from the US Antarctic Marine Living Resources Program.[1]

10.10 RESULTS

10.10.1 Krill, salps, and sea ice

Previous reports on the historical (1977–2001) Elephant Island area data sets indicated a significant decrease in krill abundance and concurrent increase in *S. thompsoni* abundance after 1989 that was associated with atmospheric warming and decreased seasonal sea ice coverage since the 1940s (Siegel and Loeb 1995; Loeb et al. 1997; Siegel et al. 1998, 2002). Updated data sets (Figure 10.8) support these earlier findings with the significance occurring between mean concentrations in 1977–1989 and 1990–2004 (Mann-Whitney U tests, $n = 21$, $P < 0.05$ for salps, $n = 22$, $P < 0.02$ for krill) associated with a significant reduction in sea ice extent between the two periods ($n = 25$, $P < 0.05$). Accordingly, krill and salps have respective positive and negative correlations with sea ice extent over the entire period (krill: $n = 22$, Kendall's $T = +0.30$, $P = 0.05$; salps: $n = 21$; $T = -0.46$, $P < 0.01$). Although sea ice extent has decreased somewhat since 1989 (Figure 10.8c) there is no significant difference in this index, krill, or salp abundance between the 1990–1998 and 1999–2004 periods. Additionally,

[1]US AMLR Program Field Season Reports, published on an annual basis from 1989 through 2002 as SWFSC Administrative Reports (1989–1999) and NOAA Technical Memoranda (2000–2005). See also a series of short papers under the general heading of US AMLR Program in the 1991–1997 review issues of the *Antarctic Journal of the US* published by the National Science Foundation. Copies available from US AMLR Program, 8604 La Jolla Shores Drive, La Jolla, CA 92037, USA.

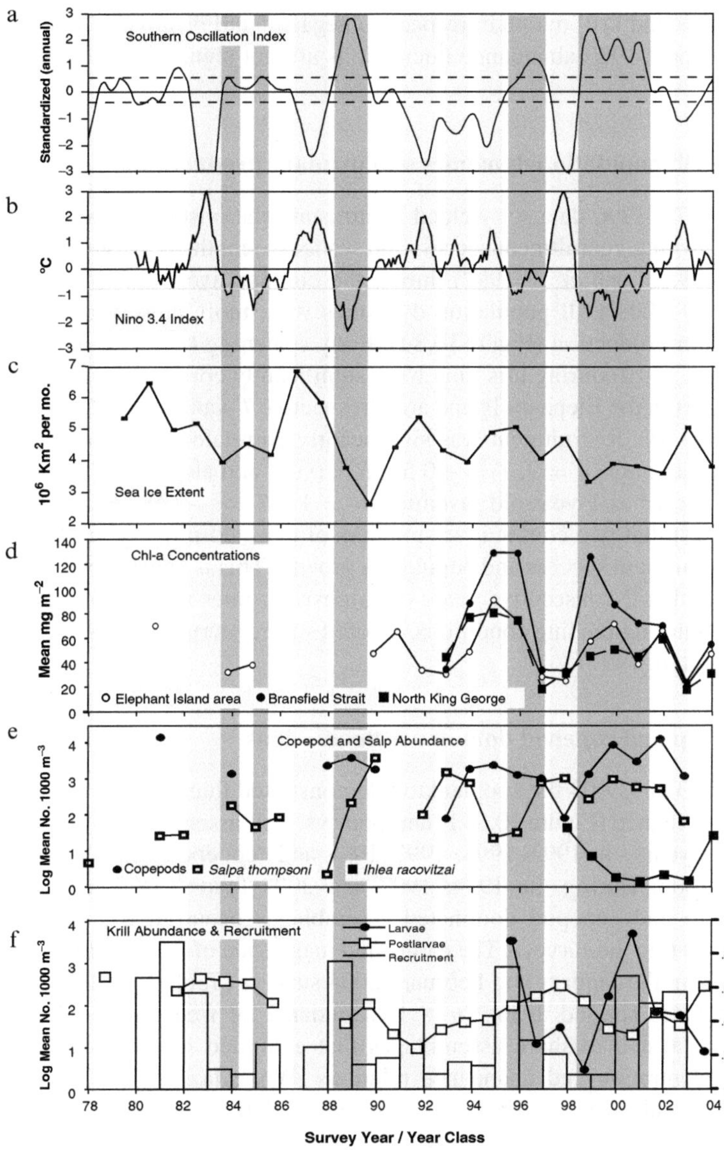

Figure 10.8. Long-term time series constructed from observations made in the Elephant Island area showing fluctuations of: (a) the 1979–2004 Southern Oscillation Index (atmospheric pressure anomaly, Tahiti-Darwin, Australia); (b) 1979–2004 Niño 3.4 Index (sea surface temperature anomaly, 5° N to 5° S, 170° W to 120° W); (c) sea ice extent; (d) integrated Chl-a (0–100 m); (e) copepod, *Salpa thompsoni* and *Ihlea racovitzai* abundance; (f) abundance of Antarctic krill post larvae, larvae and krill recruitment (solid line, R1). Light and dark grey bars represent, respectively, strong El Niño and La Niña episodes based on a threshold of ±0.65°C for the Oceanic Niño Index (3-month running mean of sea surface temperature anomalies in the Niño 3.4 region). The sea ice index is a numerical integration of the area under a seasonal curve of spatial ice cover and indexes the spatial and temporal extent of sea ice (Hewitt 1997).

although salps and krill maintain respective negative and positive associations with sea ice their post-1989 abundance fluctuations are not significantly correlated with its extent.

10.10.2 Krill population dynamics and primary production

While the 1977–1996 data set yielded significant relationships between early krill seasonal spawning, recruitment success and sea ice extent the previous winter (Siegel and Loeb 1995; Loeb et al. 1997) during the comparatively stable sea ice conditions of 1990–2004 krill population dynamics were more closely associated with phytoplankton production (Figure 10.8d). Early spawning ($n = 15$) and larval krill abundance ($n = 10$) during this time were significantly correlated with mean Chl-a concentrations in the Elephant Island area (respective T values +0.62, $P < 0.01$ and +0.51, $P < 0.05$). Recruitment was significantly correlated with larval abundance the previous summer ($n = 9$, $T = +0.54$, $P < 0.05$) and also with Chl-a concentrations encountered as 1-year-old juveniles ($n = 15$, $T = +0.48$, $P = 0.01$). These results suggest that two consecutive summers of elevated primary production favor elevated recruitment success and population growth. This is consistent with previous observations that 2 consecutive years of extensive winter sea ice, and presumably good overwintering feeding conditions, favored strong recruitment (Siegel and Loeb 1995; Loeb et al. 1997).

10.10.3 Salp and copepod dominance fluctuations

Between 1993 and 2004 the zooplankton demonstrated fluctuations between salp or copepod dominance (Figure 10.8e,f) and poor vs. rich assemblages. These included strong "salp years" in 1993–1994, 1997–1998 and "copepod years" in 1995–1996 and 1999–2002. During the 1994, 1997 and 2004 field seasons dramatic shifts between salp and copepod dominated assemblages occurred over 4- to 6-week periods separating the surveys. The nature and magnitude of these shifts are indicated by results from the January and February 2004 surveys (Table 10.1; Figure 10.9a,b). Cluster analyses applied to log ($n + 1$) abundance of frequent zooplankton taxa showed coastal and offshore assemblages that exhibited major changes in distribution, concentration and abundance relations. While taxa of the coastal cluster exhibited typical two- to three-fold seasonal variations in abundance, changes were particularly extreme offshore where a broadly distributed, depauperate, salp-dominated assemblage was supplanted by a spatially restricted, copepod-dominated assemblage. Here mean and median abundance of total zooplankton and dominant taxa (copepod species *Calanoides acutus, Rhincalanus gigas, Calanus propinquus* and *Metridia gerlachei*; chaetognaths; larvae of the euphausiid *Thysanoessa macrura*; pteropod *Limacina helicina*) were one to two orders of magnitude, and significantly greater (ANOVA, $P < 0.01$), than the previous survey.

Table 10.1. Taxonomic composition of zooplankton clusters during (A) January and (B) February–March, 2004. R and % are rank and proportions of total abundance (N/1000m^3) represented by each taxon. (L) indicate larval stages. Asterisks denote significant abundance differences between the two clusters based on ANOVA: *** P < 0.001; **P < 0.01; *P < 0.05.

	JANUARY 2004											FEBRUARY–MARCH 2004											
	Cluster 1 (Coastal) N = 44						Cluster 2 (Offshore) N = 47					Cluster 1 (Coastal) N = 82						Cluster 2 (Offshore) N = 15					
Taxa	*R*	%	MEAN		STD	MED	*R*	%	MEAN	STD	MED	*R*	%	MEAN		STD	MED	R	%	MEAN		STD	MED
Salps:																							
Salpa thompsoni	2	10.3	143.0		187.4	63.8	1	31.0	212.8	344.3	105.5	8	3.3	98.9		166.2	30.3		1.1	257.6	*	437.2	91.5
Ihlea racovitzai	6	5.3	74.4	**	148.7	12.4		0.3	2.0	8.6	0.0		0.9	26.3	*	50.5	1.3		–	–		–	–
Copepods:																							
Calanoides acutus	3	10.1	140.4		189.5	75.7	3	13.4	91.8	167.6	35.2	2	9.4	286.7		208.3	227.5	1	28.1	6468.4	***	12836.5	1078.2
Calanus propinquus	5	6.7	93.3	***	78.3	62.5	4	6.6	45.4	42.4	38.4	3	9.1	278.0		248.2	204.0	3	15.2	3492.9	***	5904.4	1286.4
Rhincalanus gigas		1.2	16.8	***	24.6	5.7		0.2	1.4	4.3	0.0	7	3.7	113.2		130.8	74.0	2	17.5	4017.9	***	8842.8	768.5
Metridia gerlachei	1	35.1	488.9	***	900.5	142.3	10	1.2	8.5	19.7	1.2	1	43.1	1311.0		3043.3	398.2	6	7.4	1702.2		2293.3	240.2
Pleuromama robusta	10	2.2	30.0	*	62.2	9.1		0.9	6.2	21.6	0.0		0.9	27.7		90.0	0.0		0.1	31.1		96.8	0.0
Pareuchaeta antarctica	9	3.7	51.7	***	88.8	18.4		0.8	5.7	12.4	0.0	6	3.8	115.1		134.6	69.0	10	1.4	332.0	**	436.6	135.7
Heterorhabdus sp.		–	–		–	–		–	–	–	–		0.1	3.7		10.7	0.0		1.2	265.1	*	954.1	0.0
Other copepods		–	–		–	–		–	–	–	–		0.9	26.1		59.2	0.7	9	1.5	333.5	***	306.9	234.6
Euphausiids:																							
Euphausia superba	7	4.5	62.3		175.7	1.4	5	4.1	28.2	68.1	3.1	9	2.7	82.7		292.6	2.5		0.1	23.7		39.8	2.1
Euphausia superba (L)		0.6	8.8		13.4	2.2		0.8	5.3	15.4	0.0	10	1.4	42.0		60.0	18.8	8	2.0	467.1	**	1263.6	107.8
Thysanoessa macrura	4	9.7	135.4		136.5	101.9	2	25.6	176.1	287.9	43.7	4	8.1	246.1	**	343.6	117.3		0.0	8.3		10.9	2.4
Thysanoessa macrura (L)		0.5	6.3		11.9	0.2	8	2.9	19.7	87.5	2.1		1.0	31.5		66.8	6.7	5	8.1	1865.3	***	1906.6	1368.8
Euphausia frigida		1.4	18.9		34.4	1.9	9	2.0	13.4	62.3	0.0		1.3	39.7		67.4	5.6		0.1	11.6		33.3	0.0
Euphausia frigida (L)		–	–		–	–		–	–	–	–		0.1	2.6		7.9	0.0		0.1	22.6	***	45.8	8.0
Euphausia triacantha		–	–		–	–		–	–	–	–		0.0	1.2		4.3	0.0		0.0	2.0		5.1	0.0
Amphipods:																							
Cyllopus lucasii		0.3	4.4	**	6.7	2.5		0.2	1.7	1.9	1.3		0.1	3.4		4.9	1.3		0.0	3.7		4.2	2.0
Cyllopus magellanicus		0.0	0.6		1.7	0.0		0.0	0.3	0.9	0.0		0.0	0.8		1.1	0.4		0.0	1.9	**	2.2	1.2
Hyperiella dilatata		0.0	0.4		1.0	0.0		0.1	0.4	0.6	0.0		0.1	1.7		8.1	0.0		0.1	13.5	*	38.6	0.8

Table 10.1. (Continued)

	JANUARY 2004											FEBRUARY–MARCH 2004												
	Cluster 1 (Coastal) N = 44						Cluster 2 (Offshore) N = 47					Cluster 1 (Coastal) N = 82						Cluster 2 (Offshore) N = 15						
Taxa	*R*	%	MEAN		STD	MED	*R*	%	MEAN	STD	MED	*R*	%	MEAN		STD	MED	R	%	MEAN		STD	MED
Amphipods:																							
Primno macropa		0.2	3.2		5.2	1.4		1.1	7.4	13.2	3.3		0.5	13.7		17.3	7.7		0.1	29.5	*	57.3	3.6
Themisto gaudichaudii		0.2	2.2		4.2	0.6		0.5	3.5	4.8	2.0		0.2	4.7		6.9	2.5		0.0	3.2		2.6	2.4
Vibilia antarctica		0.1	0.9		1.3	0.4		0.1	0.5	0.8	0.0		0.0	0.9		2.5	0.0		0.0	1.3		2.8	0.0
Pteropods:																							
Clione limacina		0.1	0.9		2.6	0.0		0.0	0.3	0.8	0.0		0.0	0.3		0.7	0.0		0.0	1.5	***	2.3	0.4
Limacina helicina		1.5	20.3		37.1	7.7	6	3.5	23.9	37.1	4.2		0.4	13.5		41.6	0.8		0.9	201.2	***	267.5	78.6
Spongiobranchaea australis		0.2	2.4		4.3	0.8		0.4	2.5	2.1	2.2		0.1	2.1		4.7	0.4		0.1	26.4	***	32.5	15.6
Others:																							
Chaetognaths	8	3.7	52.1	*	82.4	18.5	7	3.1	21.1	33.1	10.4	5	7.2	218.5		239.9	113.3	7	4.2	957.3	*	1839.9	291.3
Radiolaria		0.3	3.6		8.4	0.5		0.5	3.6	7.4	0.8		0.1	3.2		7.6	0.0	4	10.1	2323.9	**	6895.4	126.2
Ostracods		2.0	27.5	**	62.7	2.1		0.4	2.6	5.5	0.4		1.0	30.0		89.0	1.0		0.0	9.7		19.0	0.0
Tomopteris spp.		0.1	2.1	**	3.3	0.4		0.1	0.7	1.2	0.4		0.0	1.1		3.0	0.0		0.0	3.9	**	6.5	0.4
Sipunculids		0.0	0.3		0.6	0.0		0.1	0.4	1.4	0.0		0.1	1.6		5.9	0.0		0.0	0.0		0.1	0.0
Diphyes antarctica		0.0	0.4		0.8	0.0		0.0	0.2	0.7	0.0		0.0	0.3	*	0.5	0.0		–	–		–	–
Electrona spp. (L)		–	–		–	–		–	–	–	–		0.0	1.1		2.9	0.0		0.0	1.9		5.5	0.0
Lepidonotothen kempi (L)		–	–		–	–		–	–	–	–		0.0	0.6		1.5	0.0		0.0	1.6		5.8	0.0
Lepidonotothen larseni (L)		0.1	0.8		1.4	0.0		0.1	0.9	2.7	0.0		0.0	0.5		2.2	0.0		–	–		–	–
TOTAL			1392.5	***	1105.2	1038.7			686.5	440.5	577.7			3041.2		3371.8	2165.8			22979.7	***	39778.2	8577.6
TAXA		24	24.3	***	3.6	24		24	19.1	3.6	20		33	20.9		3.4	21		31	22.1		2.2	21

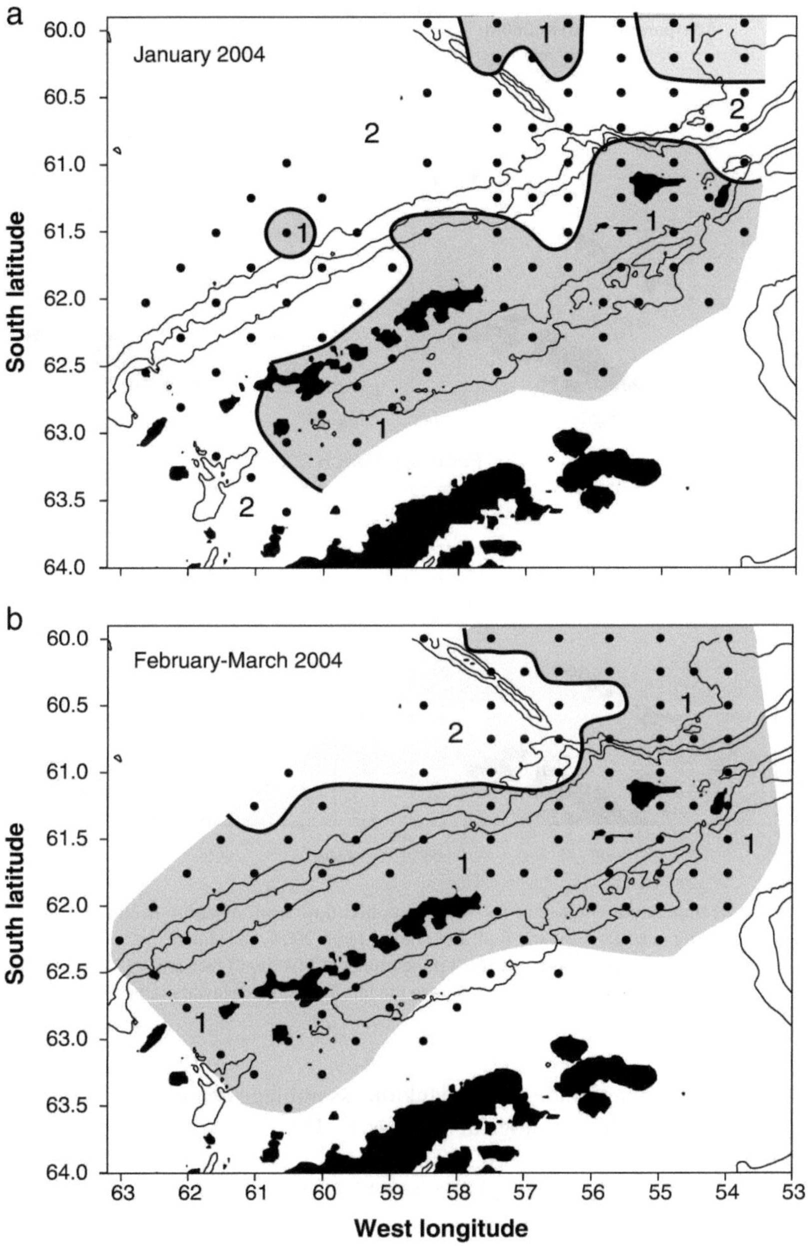

Figure 10.9. Distribution of two zooplankton clusters described in the Table 10.1 during (a) January and (b) February–March, 2004.

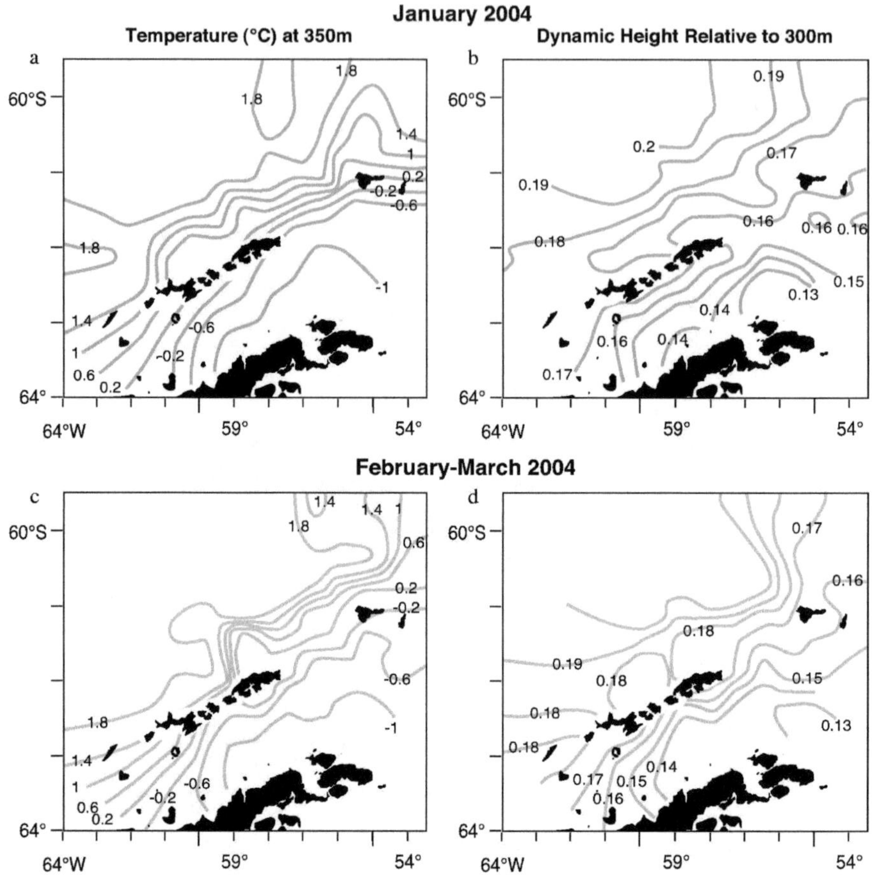

Figure 10.10. Temperature isolines at 350 m and implied flow from dynamic height at 300 m relative to the surface during (a,b) January and (c,d) February–March 2004. Variability of the 1.8°C isotherm at 350 m is reflective of changes in the sACCf. Temperatures at this depth that are <1.0 C indicate waters that are not circumpolar in origin. Implied flow is to the left from high to low dynamic height values (e.g., the northeast).

The rapidity and magnitude of zooplankton assemblage change suggest underlying hydrographic rather than biological processes. This is supported by coincidental changes in the position of the sACCf indicated by the location of 1.8° C water at depth (Orsi et al. 1995; Figure 10.10a,b) which shows a more poleward location, especially north of Livingston and Elephant Islands, during February–March. The corresponding dynamic height distributions (Figure 10.10c,d) show decreased flow out of Bransfield Strait between King George and Elephant Islands and increased eastward flow along the southern island shelves in Bransfield Strait and outer island shelves in Drake Passage in February–March. Distribution patterns of *S. thompsoni* and high latitude salp species *Ihlea racovitzai*, a marker for Weddell Sea Water (Foxton 1971), indicate northeastward flow within the upper water column during February–March. A significant phytoplankton bloom occurred southeast of King George Island during February–March.

Hydrographic and ecological conditions represented during the 1998 "salp year" and 1999 "copepod year" (Loeb et al., submitted) were consistent with those observed during 2004: increased primary production, particularly in Bransfield Strait; greatly increased zooplankton abundance offshore; intensified northeast flow associated with poleward movement of the sACCf and/or Bndy; decreased Weddell Sea water influence.

10.10.4 Salp source areas

Salpa thompsoni is generally associated with ACC waters north of the sACCf (Pakhomov and McQuaid 1996; Nicol et al. 2000; Pakhomov et al. 2002) and historically noted as rarely found at stations influenced by Weddell Sea water (Schnack-Schiel and Mujica 1991). However, between 1993 and 2001 concentrations of *S. thompsoni* at stations with Weddell Sea water temperature-salinity characteristics were significantly greater than those with ACC water characteristics during 8 of 16 surveys (ANOVA's, $P \leq 0.01$ in all cases) indicating substantial input from the east and/or southeast of Elephant Island. In contrast, during 2002–2004 *S. thompsoni* concentrations associated with offshore ACC water were significantly greater than those of Weddell Sea and/or mixed Bransfield Strait waters in six of eight 8 surveys (ANOVAs, $P < 0.05$) suggesting decreased importance of the Weddell Sea vs. ACC as a source area between the two periods.

10.11 DISCUSSION

10.11.1 Krill, salps, and sea ice

The Elephant Island data set reveals ecological variations on scales ranging from weeks to decades. Increased salp and decreased krill abundance here are associated with a long-term (1979–2002) trend in decreased sea ice concentrations in the Bellingshausen-western Weddell Sea sectors of the Southern Ocean (Figure 10.8c; Zwally et al. 2002; Liu et al. 2004). The significant abundance changes center around 1989 which follows a 3-year period of record sea ice decrease due to more southerly surface winds and peak surface air temperatures along the West Antarctic Peninsula (Zwally et al. 2002). The dramatic krill stock decline reflects the negative demographic effects of 3- to 4-year periods between successful krill year classes (Figure 10.8a; Priddle et al. 1988; Siegel et al. 2002). Based on the impact of the successful 1999/2000, 2000/01 and 2001/02 year classes (Figure 10.8a) it is likely that krill populations could build to earlier sizes given 2- to 3-year intervals of good recruitment in the apparent 4- to 5-year cycle (Priddle et al. 1988).

Over the last century, a poleward shift in salp distribution has occurred at a circumpolar scale, resulting in increased densities now south of the Bndy (Atkinson et al. 2004). This latitudinal shift is not directly related to sea ice extent, as decreased sea ice in the Bellingshausen and western Weddell Sea sectors since 1979 contrasts with increased extent elsewhere, especially in the Ross Sea (Zwally et al. 2002), and the post-1989 independence of salp abundance and sea ice extent in the AMLR survey area now conforms to results from other areas (Atkinson et al. 2004). This southward abundance shift coincides with increased warming of the upper and mid-depth waters of the Southern Ocean and southward migration by about 50 km of the ACC between the 1950s and mid-1990s (Levitus et al. 2000; Gille 2002). Increased salp densities are now located in coastal and offshore areas to the west (Bellingshausen Sea and west Antarctic Peninsula) and east (eastern Weddell Sea) of the Elephant Island area (Atkinson et al. 2004).

10.11.2 Salp and copepod dominance fluctuations

A variable boundary between rich and poor zooplankton zones, similar to that observed in the long-term AMLR data set (Table 10.1; Figure 10.9a,b), was apparent during the 1928–1931 *Discovery* Expeditions. Mackintosh (1934) described a "thin" zooplankton assemblage, characteristic of the East Wind Drift that, aside from krill, generally had few organisms except for occasional dense patches of the copepods *Metridia gerlachei* and *Calanoides acutus* and *S. thompsoni*. This contrasted with an abundant zooplankton assemblage in Drake Passage (Westwind Drift) dominated by the copepods *Rhincalanus gigas, Calanus propinquus* and *Calanoides acutus*, chaetognaths, and pteropod *Limacina* sp.. The abrupt change between these assemblages varied from 150 miles offshore in 1929 and 1931 to within 30 miles, and along the shelf, of the South Shetland Islands in 1930. This movement represents a latitudinal shift of about 2° that corresponds to the variability in location of the sACCf off King George Island during 1996–2000 (Sprintall 2003; Figure 10.11).

BIOMASS FIBEX and SIBEX surveys also documented shifts between rich, copepod dominated zooplankton assemblages adjacent to the South Shetland shelf and elevated Chl-a concentrations in Bransfield Strait in summer 1981 and depauperate, salp dominated assemblages and low Chl-a concentrations in 1983/84 (Jadzewski et al. 1982; Witek et al. 1985; Bodungen et al. 1986; Priddle et al. 1991; Schnack-Schiel and Mujica 1991). These changes, as well as fluctuations in sea ice development, krill distribution and reproductive success of krill predators, were believed related to atmospherically driven changes in circulation patterns in Drake Passage, Bransfield Strait and Weddell Sea that modulated flow into Bransfield Strait from Drake Passage (CDW), Gerlache Strait and Weddell Sea (Jadzewski et al. 1982; Witek et al. 1985; Priddle et al. 1988). Various authors (Priddle et al. 1988; Sahrhage 1988) attributed such changes to latitudinal movement of fronts within the ACC and hypothesized that they resulted from forcing by El Niño Southern Oscillation (ENSO) variations in the equatorial Pacific.

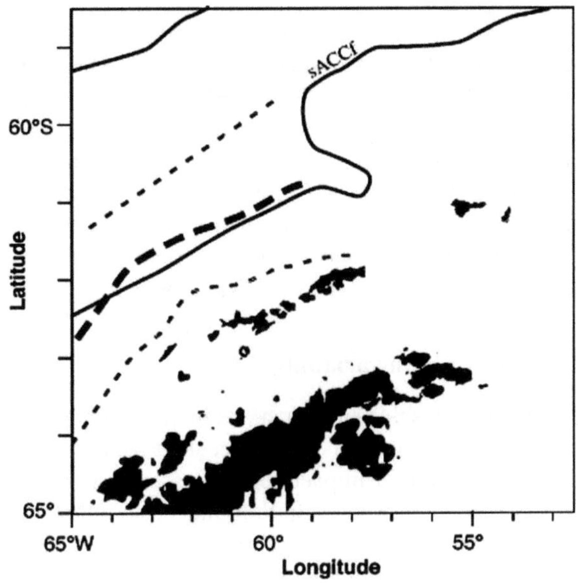

Figure 10.11. The mean position (heavy dashed line) and one standard deviation (light dashed lines) of the Southern Antarctic Circumpolar Front presented by Orsi et al. 1995 (solid line) in Drake Passage, 1996–2001. Derived from Sprintall (2003).

10.11.3 Salp source areas

The apparent Weddell Sea source of *S. thompsoni* during 1993–2001 counters previous observations that this species is restricted to warmer, ice free waters. However, like the ACC, Weddell Sea waters have warmed in recent years (Robertson et al. 2002; Fahrbach et al. 2004) and thus could provide a more favorable habitat. Additionally, poleward displacement of salps to locations as far as 68°S has been associated with warmer (>0°C) water layers upwelled and/or advected from the CDW (Pakhomov 2004). A warm core region within the westward flowing southern limb of the Weddell gyre results from injection of CDW from the ACC east of the Weddell Sea. Warming of these core waters from 1992 to 1998 followed intensified intrusion of CDW during the late 1980s and early 1990s and was associated with poleward movement of the ACC during that time (Robertson et al. 2002; Fahrbach et al. 2004). These conditions could favor the transport of salps from source areas east of the Weddell Sea into eastern Bransfield Strait.

Historically, *I. racovitzai* was reported to be essentially absent from the South Pacific Ocean sector of the Southern Ocean (60°W to 120°E) and had a distribution pattern in surface waters of the South Atlantic and South Indian Oceans similar to that of krill (Figure 10.5; Foxton 1971). Large numbers of this salp in the Antarctic Peninsula region were first noted during summer 1986 (Esnal and Daponte 1990) when they were collected from the western Weddell Sea (Joinville Island), through Bransfield Strait and off the West Antarctic Peninsula. This species was also abundant in the South Shetland Island Area during December 1990–January 1991 (Nishikawa et al. 1995). It was first noted in AMLR samples during 1998 when it comprised the fourth most abundant taxon overall. Greatest concentrations then extended across the inner northern shelves of Elephant and King George Islands and in Bransfield Strait south of Clarence and King George Islands. Elevated concentrations of *I. racovitzai* advected into the region in recent years, like those of *S. thompsoni*, are likely to be a consequence of ocean warming with a poleward habitat shift into areas influenced by the East Wind Drift.

10.11.4 ENSO and ecosystem variability

Between 1977 and 2004 krill, salps and copepods in the Elephant Island area demonstrated variations in distribution and abundance that generally conform to three phases: 1977–1989 dominated by impacts of decreased sea ice extent; 1990–1998 characterized by oscillations between salp dominated, zooplankton-poor and copepod dominated, zooplankton-rich periods; a relatively stable and productive 6-year period marked by elevated zooplankton concentrations and good krill recruitment success over, respectively, 4 and 3 consecutive years (Figures 10.8 and 10.11). Underlying each of these phases however, is a 3- to 4-year periodicity of seasonal sea ice cycles, salp-copepod abundance fluctuations and krill recruitment success that strongly suggest a dominant influence by ENSO variability (McPhaden, Chapter 1, this volume).

Investigations over the last 20 years have established that the climate system in the eastern Pacific and western Atlantic sectors of the Southern Ocean is strongly impacted by meridional atmosphere teleconnections instigated in the western tropical Pacific Ocean by ENSO variability (Karoly 1989; Cai and Baines 2001; White et al. 2002; Carleton 2003). This external forcing drives variations of sea level pressure, the Pacific South Atlantic (PSA) pattern, and corresponding Antarctic dipole pattern of sea surface temperature in high latitudes in phase with ENSO (Karoly 1989; Yuan and Martinson 2001; Liu et al. 2002). This ENSO signal is subsequently propagated eastward around the remainder of the Southern Ocean (White et al. 2002). These patterns of basin-scale climate variability influence regional climatic conditions in the vicinity of the West Antarctic Peninsula, with cool SST anomalies, weak northwesterly wind anomalies and retracted sea ice extent fluctuating in phase with tropical El Niño and warm SST anomalies, strong northwesterly wind anomalies and expanded sea ice extent fluctuating in phase with tropical La Niña (Gloerson and White 2001).

Statistically significant correlations between concentrations of larval and postlarval krill, salps, Chl-a, sea ice extent, krill recruitment success and monthly values of the atmospheric Southern Oscillation (SOI) and oceanographic Niño 3.4 indices (Figure 10.8) show that ENSO-driven variability is dominant in structuring the marine ecosystem in the west Antarctic Peninsula region (Loeb et al.

submitted). As observed in 2004, interannual and shorter-term variability in Chl-a concentrations and zooplankton composition and abundance here are associated with latitudinal movements of the sACCf and Bndy west of the Shackleton Fracture Zone: poleward movement brings the copepod-rich oceanic zooplankton assemblage and promotes elevated primary production; equatorward movement results in low primary production, favorable for *S. thompsoni* (Nicol et al. 2000; Atkinson et al. 2004; Pakhomov 2004), and elevated concentrations of *I. racovitzai*. These ecological changes suggest that latitudinal shifts in the sACCF and Bndy may coincide with changes in coastal circulation, mixing, mixed layer depth and stability as well as sea ice extent and contentration and prevailing wind conditions. Interannual variability of these environmental conditions associated with relative intensity of ACC vs. Weddell gyre influences are consistent with ENSO and the Antarctic dipole (Yuan and Martinson 2001; Liu et al. 2002; Martinson and Iannuzzi 2003).

Over the last century ENSO cycles have ranged from 2- to 7-years but typically exhibited a 3- to 5-year periodicity. However, these cycles have co-occurred with longer-term climatic "regimes", the Pacific Decadal Oscillation (PDO), characterized by 20- to 30-year periods of persistent warm or cool water temperature anomalies. The climatic expressions of the PDO are apparent at high latitudes, they prevail in both the Northern and Southern Hemispheres and are known to have significant impacts on marine ecosystems (Mantua and Hare 2002; McPhaden this volume). Cool La Niña-like PDO regimes dominated from 1890–1924 and 1947–1976, warm El Niño-like PDO regimes prevailed from 1925–1946 and 1977 to 1998, after which relatively neutral conditions have persisted (Goericke et al. 2004). As a consequence, most of our knowledge of ecosystem dynamics in the Antarctic Peninsula region has been derived from warm regimes. The long-term data sets presented here (Figures 10.8 and 10.11) coincide with initiation of the recent warm regime and extend 6 years beyond an hypothesized regime shift following the intense 1998 El Niño (Bond et al. 2003; Peterson and Schwing 2003) which may in part explain the three apparent phases of ecosystem variability: the 1977–1989 period dominated by impacts of decreased sea ice extent due to combined effects of intensified El Niño events and increased temperatures attributed to global warming; 1990–1998 period characterized by ENSO driven oscillations between salp and copepod dominance under relatively stable sea ice conditions; post-1998 period of predominantly cool or neutral ENSO conditions favorable to increased primary and secondary productivity and good krill recruitment success.

Paleo-climate records suggest the influence of ENSO-related climate variability in the Drake Passage vicinity over the last 130,000 years (Fischer et al. 2004; Turney et al. 2004). An implication of this is the food web of the Southern Ocean has evolved to adapt to ENSO variability. In particular, the 5- to 8-year longevity of krill, long relative to other euphausiid species, guarantees at least one successful recruitment season (every 3- to 7-years) in an environment dominated by ENSO-related variability (Fraser and Hofmann 2003).

10.12 ACKNOWLEDGMENTS

This work was supported by the National Oceanic and Atmospheric Administration US Antarctic Marine Living Resources Program through NOAA contract no. AB5CN0003. I thank AMLR Director Rennie Holt for facilitating this effort and Roger Hewitt for providing sea ice indices. I also thank the captains, crew, colleagues and multitudes of shipboard assistants over the years that helped develop the long-term AMLR data set. Special thanks go to A. Amos, O. Holm-Hansen and C. Hewes, for their hard work that ensured collection of high quality data sets over the years. The author is greatly appreciative of the editorial efforts by Eileen Hofmann that focused the manuscript and made it much more coherent and readable. The views expressed herein are those of the author and do not necessarily reflect the views of NOAA or any of its subagencies.

10.13 LITERATURE CITED

Ainley, D.G., C.T. Tynan, and I. Stirling. 2003. Sea ice: a critical habitat for polar marine mammals and birds. In: *Sea ice: an introduction to its physics, chemistry, biology and geology* (eds., D.N. Thomas and G.S. Dieckmann), pp. 240–266. Blackwell Science, Oxford.

Amos, A.F. 1984. Distribution of krill (*Euphausia superba*) and the hydrography of the Southern Ocean: Large-scale processes. *Journal of Crustacean Biology* 4 (Special No. 1): 306–329.

Amos, A.F. 2001. A decade of oceanographic variability in summertime near Elephant Island, Antarctica. *Journal of Geophysical Research* 106 C10: 22401–22423.

Atkinson, A., V. Siegel, E. Pakhomov, and P. Rothery. 2004. Long-term decline in krill stock and increase in salps within the Southern Ocean. *Nature* 432: 100–103.

Bodungen, B.V., V.S. Smetacek, M.M. Tilzer, and B. Zeitzschel. 1986. Primary production and sedimentation during spring in the Antarctic Peninsula region. *Deep-Sea Research* 33: 177–194.

Bond, N.A., J.E. Overland, M. Spillane, and P. Stabeno. 2003. Recent shifts in the state of the North Pacific. *Geophysical Research Letters* 30: Art. No. 2183.

Brinton, E. 1991. Distribution and population structures of immature and adult *Euphausia superba* in the western Bransfield Strait region during the 1986–87 summer. *Deep-Sea Research* 38: 1169–1193.

Cai, W. and P.G. Baines. 2001. Forcing of the Antarctic Circumpolar Wave by El Niño–Southern Oscillation teleconnections. *Journal of Geophysical Research* 106: 9019–9038.

Capella, J.E., R.M. Ross, L.B. Quetin, and E.E. Hofmann. 1992. A note on the thermal structure of the upper ocean in the Bransfield Strait–South Shetland Islands region. *Deep-Sea Research* 39: 1221–1229.

Carleton, A.M. 2003. Atmospheric teleconnections involving the Southern Ocean. *Journal of Geophysical Research* 108: Art. No. 8080.

Casereto, B.E. and T. Nemoto. 1986. Salps of the Southern Ocean (Australian Sector) during the 1983–1984 summer, with special reference to the species *Salpa thompsoni*, Foxton 1961. *Memoirs of the National Institute of Polar Research* 40 (Special Issue): 221–239.

Clarke, A. 1990. Temperature and evolution: Southern Ocean cooling and the Antarctic marine fauna. In: *Antarctic ecosystems. Ecological change and conservation* (eds., K.R. Kerry and G. Hempel), pp. 9–22. Springer, Berlin.

Clarke, A., D.K.A. Barnes, and D.A. Hodgson. 2005. How isolated is Antarctica? *Trends in Ecology and Evolution* 20: 1–3.

Clowes, A.J. 1934. Hydrology of the Bransfield Strait. *Discovery Report* 9: 1–64.

Comiso, J.C. 2003. Large-scale characteristics and variability of the global sea ice cover. In: *Sea ice: an introduction to its physics, chemistry, biology and geology* (eds., D.N. Thomas and G.S. Dieckmann), pp. 112–142. Blackwell Science, Oxford.

Daponte, M.C., F.L. Capitanio, and G.B. Esnal. 2001. A mechanism for swarming in the tunicate *Salpa thompsoni*. *Antarctic Science* 13: 240–245.

de Baar, H.J.W., J.R.M. de Jong, D.C.E. Bakker, B.M. Löscher, C. Veth, U. Bathmann, and V. Smetacek. 1995. Importance of iron for plankton blooms and carbon dioxide drawdown in the Southern Ocean. *Nature* 373: 412–415.

Dinniman, M.S., J.M. Klinck, and W.O Smith. 2003. Cross-shelf exchange in a model of the Ross Sea circulation and biogeochemistry. Deep-Sea Research II Topical Studies in Oceanography 50: 3103–3120.

Eastman, J.T. 1993. *Antarctic fish biology: evolution in a unique environment*, 322 pp. Academic Press, New York.

Eicken, H. 1992. The role of sea ice in structuring Antarctic ecosystems. *Polar Biology* 2: 3–13.

El Sayed, S.Z. 1985. Plankton of the Antarctic Seas. In: *Key environments Antarctica* (eds., W.N. Bonner and D.W.H. Walton). pp. 135–153. Pergamon, Oxford.

Esnal, G.B. and M.C. Daponte. 1990. Stolon development and some aspects of musculature in the characterization of *Ihlea racovitzai* (van Beneden, 1913) and *Ihlea magalhanica* (Apstein, 1894) (Tunicata, Thaliacea). *Polar Biology* 10: 265–268.

Fach, B.A., E.E. Hofmann, and E.J. Murphy. 2002. Modeling studies of Antarctic krill *Euphausia superba* survival during transport across the Scotia Sea. *Marine Ecology Progress Series* 231: 187–203.

Fahrbach, E., M. Hoppema, G. Rohardt, M. Schröder, and A. Wisotzki. 2004. Decadal-scale variations of water mass properties in the deep Weddell Sea. *Ocean Dynamics* 54: 77–91.

Fischer, H., F. Traufetter, H. Oerter, R. Weller, and H. Miller. 2004. Prevalence of the Antarctic Circumpolar Wave over the last two millennia recorded in Dronning Maud Land ice. *Geophysical Research Letters* 31: Art. No. L08202.

Foxton, P. 1966. The distribution and life history of *Salpa thompsoni* Foxton, with observations on a related species, *Salpa gerlachei*, Foxton. *Discovery Report* 34: 1–116.

Foxton, P. 1971. On *Ihlea magalhanica* (Apstein) (Tunicata: Salpidae) and *Ihlea racovitzai* (Van Beneden). *Discovery Report* 35: 1–6.

Fraser, F.C. 1936. On the development and distribution of the young stages of krill (*Euphausia superba*). *Discovery Report* 14: 1–192.

Fraser, W.R. and E.E. Hofmann. 2003. A predator's perspective on causal links between climate change, physical forcing and ecosystem response. *Marine Ecology Progress Series* 265: 1–15.

Garcia, M.A., O. López, J. Sospedra, M. Espino, V. Gracia, G. Morrison, P. Rojas, J. Figa, J. Puigdefabregas, and A.S. Arcilla. 1994. Mesoscale variability in the Bransfield Strait region (Antarctica) during austral summer. *Annales Geophysicae. Atmospheres, hydrospheres and space sciences*12: 856–867.

Gille, S.T. 2002. Warming of the Southern Ocean since the 1950's. *Science* 295: 1275–1277.

Gloerson, P. and W.B. White. 2001. Reestablishing the circumpolar wave in sea ice around Antarctica from one winter to the next. *Journal of Geophysical Research* 106: 4391–4395.

Goericke, R. et al. 2004. The state of the California Current, 2003–2004: a rare "normal" year. *California Cooperative Oceanic Fisheries Investigations Reports* 45: 27–59.

Hempel, G. 1985. On the biology of polar seas, particularly the Southern Ocean. In: *Marine biology of polar regions and effects of stress on marine organisms* (eds., J.S. Gray and M.E. Christiansen), pp. 3–33. Wiley, Chichester, UK.

Hewitt, R.P. 1997. Areal and seasonal extent of sea-ice cover off the northwestern side of the Antarctic Peninsula: 1979 to 1996. *CCAMLR Science* 4: 65–73.

Hofmann, E.E., J.E. Capella, R.M. Ross, and L.B. Quetin. 1992. Models of the early life history of *Euphausia superba* - Part 1. Time and temperature dependence during the descent-ascent cycle. *Deep-Sea Research* 39: 1177–1200.

Hofmann, E.E., J.M. Klinck, C.M. Lascara, and D.A. Smith. 1996. Water mass distribution and circulation west of the Antarctic Peninsula and including Bransfield Strait. In: *Foundations for Ecological Research West of the Antarctic Peninsula. Antarctic Research* Series 70 (eds., R.M. Ross, E.E. Hofmann, and L.B. Quetin), pp. 61–80. American Geophysical Union, Washington, DC.

Hofmann, E.E. and C.M. Lascara. 2000. Modeling the growth dynamics of Antarctic krill *Euphausia superba*. *Marine Ecology Progress Series* 194: 219–231.

Hofmann, E.E. and Y.S. Hüsrevoglu. 2003. A circumpolar modeling study of habitat control of Antarctic krill (*Euphausia superba*) reproductive success. *Deep-Sea Research II* 50: 3121–3142.

Hofmann, E.E. and E.J. Murphy. 2004. Advection, krill and Antarctic marine ecosystems. *Antarctic Science* 16: 487–499.

Holm-Hansen, O. A.F. Amos, N. Silva S., V. Villafañe, and E.W. Helbling. 1994. *In situ* evidence for a nutrient limitation of phytoplankton growth in pelagic Antarctic waters. *Antarctic Science* 6: 315–324.

Holm-Hansen, O., C.D. Hewes, V.E. Villafañe, E.W. Helbling, N. Silva, and T. Amos. 1997. Distribution of phytoplankton and nutrients in relation to different water masses in the area around Elephant Island, Antarctica. *Polar Biology* 18: 145–153.

Holm-Hansen, O., M. Naganobu, S. Kawaguchi, T. Kameda, I. Krasovski, P. Tchernyshkov, J. Priddle, R. Korb, M. Brandon, D. Demer, R.P. Hewitt, M. Kahru, and C.D. Hewes. 2004. Factors influencing the distribution, biomass, and productivity of phytoplankton in the Scotia Sea and adjoining waters. *Deep-Sea Research II* 51: 1333–1350.

Huntley, M., D.M. Karl, P. Niiler, and O. Holm-Hansen. 1991. Research on Antarctic Coastal Ecosystem Rates (RACER): an interdisciplinary field experiment. *Deep-Sea Research* 38: 91–941.

Huntley, M. and E. Brinton. 1991. Mesoscale variation in growth and early development of *Euphausia superba* Dana in the western Bransfield Strait region. *Deep-Sea Research* 38: 1213–1240.

Ichii, T., K. Katayama, N. Obitsu, H. Ishii, and M. Naganobu. 1998. Occurrence of Antarctic krill (*Euphausia superba*) concentrations in the vicinity of the South Shetland Islands: relationship to environmental parameters. *Deep-Sea Research I* 45: 1235–1262.

Jadzewski, K., W. Kittel, and K. Lotocki. 1982. Zooplankton studies in the southern Drake Passage and in the Bransfield Strait during the austral summer (BIOMASS-FIBEX, February–March 1981). *Polish Polar Research* 3: 203–242.

Karoly, D.J. 1989. Southern hemisphere circulation features associated with El Niño–Southern Oscillation events. *Journal of Climate* 2: 1239–1252.

Kawaguchi, K., S. Ishikawa, and O. Matsuda. 1986. The overwintering of Antarctic krill (*Euphausia superba*) under the coastal fast ice off the Ongul Islands in Lutzow-Holm Bay, Antarctica. *Memoirs of National Institute of Polar Research* 44 (Special Issue): 67–85.

Knox, G.A. 1994. *The biology of the Southern Ocean. Studies in polar research*, 444 pp. Cambridge University Press, Cambridge.

Lancraft, T.M., J.J. Torres, and T.L. Hopkins. 1989. Micronekton and macrozooplankton in the open waters near Antarctic ice edge zones (AMERIEZ 1983 and 1986). *Polar Biology* 9: 225–233.

Lancraft, T.M., T.L. Hopkins, J.H. Torres, and J. Donnelly. 1991. Oceanic micronektonic/ macrozooplanktonic community structure and feeding in ice covered Antarctic waters during the winter (AMERIEZ 1988). *Polar Biology* 11: 157–167.

Lascara C.M., E.E. Hofmann, R.M. Ross, and L.B. Quetin. 1999. Seasonal variability in the distribution of Antarctic krill, *Euphausia superba*, west of the Antarctic Peninsula. *Deep-Sea Research I* 46: 951–984.

Laws, R.M. 1985. The ecology of the Southern Ocean. *American Scientist* 73: 26–40.

Lawver, L.A. and L.M. Gahagan. 2003. Evolution of Cenozoic seaways in the circum-Antarctic region. *Paleogeography, Palaeoclimatology, Palaeoecology* 198: 11–37.

Levitus, S., J.I. Antonov, T.P. Boyer, and C. Stephens. 2000. Warming of the world ocean. *Science* 287: 2225–2229.

Liu, J., X. Yuan, D. Rind, and D.G. Martinson. 2002. Mechanism study of the ENSO and southern high latitude climate teleconnections. *Geophysical Research Letters* 29: Art. No. 1679.

Liu, J., J.A. Curry, and D.G. Martinson. 2004. Interpretation of recent Antarctic sea ice variability. *Geophysical Research Letters* 31: Art. No. L02205.

Loeb, V., V. Siegel, O. Holm-Hansen, R. Hewitt, W. Fraser, W. Trivelpiece, and S. Trivelpiece. 1997. Effects of sea-ice extent and krill or salp dominance on the Antarctic food web. *Nature* 387: 897–900.

Loeb, V.J., E.E. Hofmann, J.M. Klinck, O. Holm-Hansen, and W.B. White. Submitted. Hydrographic Control of the Marine Ecosystem in the South Shetland-Elephant Island and Bransfield Strait Region. *Deep-Sea Research.*

Longhurst, A.R. 1976. Vertical migration. In: *The ecology of the seas* (eds., D.H. Cushing and J.J. Walsh), pp. 116–137. Blackwell Scientific, Oxford.

Lutjeharms, J.R.E., N.W. Walters, and B.R. Allanson. 1985. Oceanic frontal systems and biological enhancement. In: *Antarctic nutrient cycles and food webs* (eds., W.R. Siegfried, P.R. Condy, and R.M. Laws), pp. 11–21. Springer, Berlin.

Mackintosh, N.A. 1934. Distribution of the macroplankton in the Atlantic sector of the Antarctic. *Discovery Report* 9: 65–160.

Makarov, R.R. and C.J.I. Denys. 1981. *Stages of sexual maturity of Euphausia superba.* BIOMASS Handbook 11.

Mantua, N.J. and S.R. Hare. 2002. The Pacific Decadal Oscillation. *Journal of Oceanography* 58: 35–44.

Martinson, D.G. and R.A. Iannuzzi. 2003. Spatial/temporal patterns in Weddell gyre characteristics and their relationship to global climate. *Journal of Geophysical Research* 108: Art. No. 8083.

Marr, J.W.S. 1962. The natural history and geography of the Antarctic krill (*Euphausia superba* Dana). *Discovery Report* 32: 33–464.

McPhaden, M.J. This Volume. El Niño and La Niña: a review of physical mechanisms, climate impacts, and ecosystems responses. In: *The impact of environmental variability on ecological systems* (eds., D.A. Vasseur and K.S. McCann). Springer, Berlin.

Murphy, E.J., S.E. Thorpe, J.L. Watkins, and R. Hewitt. 2004a. Modeling the krill transport pathways in the Scotia Sea: spatial and environmental connections generating the seasonal distribution of krill. *Deep-Sea Research II* 51: 1435–1456.

Murphy, E.J., J.L. Watkins, M.P. Meredith, P. Ward, P.N. Trathan, and S.E. Thorpe. 2004b. South Antarctic Circumpolar Current Front to the northeast of South Georgia: horizontal advection of krill and its role in the ecosystem. *Journal of Geophysical Research – Oceans* 109: Art. No. C01029.

Nicol, S., T. Pauly, N.L. Bindoff, S. Wright, D. Thiele, G.W. Hosie, P.G. Strutton, and E. Woehler. 2000. Ocean circulation off east Antarctica affects ecosystem structure and sea-ice extent. *Nature* 406: 504–507.

Niiler, P.P., A. Amos, and J.-H. Hu. 1991. Water masses and 200m relative geostrophic circulation in the western Bransfield Strait region. *Deep-Sea Research* 38: 943–959.

Nishikawa, J., M. Naganobu, T. Ichii, H. Ichii, M. Terazaki, and K. Kawaguchi. 1995. Distribution of salps near the South Shetland Islands during austral summer, 1990–1991, with special reference to krill distribution. *Polar Biology* 15: 31–39.

Orsi, A.H., T. Whitworth III, and V.D. Nowlin Jr. 1995. On the meridional extent and fronts of the Antarctic Circumpolar Current. *Deep-Sea Research I* 42: 641–673.

Pakhomov, E.A. and C.D. McQuaid. 1996. Distribution of surface zooplankton and seabirds across the Southern Ocean. *Polar Biology* 16: 271–286.

Pakhomov, E.A., P.W. Froneman, and R. Perissinotto. 2002. Salp/krill interactions in the Southern Ocean: spatial segregation and implications for the carbon flux. *Deep-Sea Research II* 49: 1881–1907.

Pakhomov, E.A. 2004. Salp/krill interactions in the eastern Atlantic sector of the Southern Ocean. *Deep-Sea Research II* 51: 2645–2660.

Patarnello, T., L. Bargelloni, V. Varotto, and B. Battaglia. 1996. Krill evolution and the Antarctic ocean currents: evidence of vicariant speciation as inferred by molecular data. *Marine Biology* 126: 603–608.

Peterson, W.T. and F.B Schwing. 2003. A new climate regime in northeast Pacific ecosystems. *Geophysical Research Letters* 30: Art. No. 1896.

Petit, J.R. et al. 1999. Climate and atmospheric history of the past 420,000 years from the Vostok ice core, Antarctica. *Nature* 399: 429–436.

Pollard, R.T., M.I. Lucan, and J.F. Read. 2002. Physical controls on biogeochemical zonation in the Southern Ocean. *Deep-Sea Research II: Topical studies in Oceanography* 49: 3289–3305.

Prézelin, B.B., E.E. Hofmann, M. Moline, and J.M. Klinck. 2004. Physical forcing of phytoplankton community structure and primary production in continental shelf waters of the Western Antarctic Peninsula. *Journal of Marine Research* 62: 419–460.

Priddle, J., J.P. Croxall, I. Everson, R.B. Heywood, E.J. Murphy, P.A. Prince, and C.B. Sear. 1988. Large-scale fluctuations in distribution and abundance of krill – a discussion of possible causes. In: *Antarctic ocean and resources variability* (ed., D. Sahrhage) pp. 169–182. Springer, Berlin.

Priddle, J., F. Brandini, M. Lipski, and M.R. Thorley. 1991. Pattern and variability of phytoplankton biomass in the Antarctic Peninsula region: an assessment of the BIOMASS cruises. In: *Southern Ocean ecology: the BIOMASS perspective* (ed., S.Z. Sayed), pp. 49–61. Cambridge University Press, Cambridge.

Quetin, L.B. and R.M. Ross. 2003. Episodic recruitment in Antarctic krill *Euphausia superba* in the Palmer LTER study region. *Marine Ecology Progress Series* 259: 185–200.

Reid, K. and J.P. Croxall. 2001. Environmental response of upper trophic-level predators reveals a system change in an Antarctic marine ecosystem. *Proceedings of the Royal Society of London, Series B* 268: 377–384.

Reid, K., E.J. Murphy, V. Loeb, and R.P. Hewitt. 2002. Krill population dynamics in the Scotia Sea: variability in growth and mortality within a single population. *Journal of Marine Systems* 36: 1–10.

Robertson, R., M. Visbeck, A.L. Gordon, and E. Fahrbach. 2002. Long-term temperature trends in the deep waters of the Weddell Sea. *Deep-Sea Research II* 49: 4791–4806.

Ross, R.M., E.E. Hofmann, and L.B. Quetin. 1996. *Foundations for ecological research west of the Antarctic Peninsula. Antarctic Research Series 70*, 448 pp. American Geophysical Union, Washington, DC.

Sahrhage, D. 1988. Some indications for environmental and krill resources variability in the Southern Ocean. In: *Antarctic ocean and resources variability* (ed., D. Sahrhage), pp. 33–40. Springer, Berlin.

Schnack-Schiel, S.B. and Mujica, A. 1991. The zooplankton of the Antarctic Peninsula region. In: *Southern Ocean ecology: the BIOMASS perspective* (ed., S.Z. Sayed), pp. 79–92. Cambridge University Press, Cambridge.

Schnack-Schiel, S.B. 2003. The macrobiology of sea ice. In: *Sea ice: an introduction to its physics, chemistry, biology and geology* (eds., D.N. Thomas and G.S. Dieckmann), pp. 211–239. Blackwell Science, Oxford.

Siegel, V. 1987. Age and growth of Antarctic Euphausiacea (Crustacea) under natural conditions. *Marine Biology* 96: 483–495.

Siegel, V. 1988. A concept of seasonal variation of krill (*Euphausia superba*) distribution and abundance west of the Antarctic Peninsula. In: *Antarctic ocean and resources variability* (ed., D. Sahrhage), pp. 219–230. Springer, Berlin.

Siegel, V. and V. Loeb. 1995. Recruitment of Antarctic krill (*Euphausia superba*) and possible causes for its variability. *Marine Ecology Progress Series* 123: 45–56.

Siegel, V., V. Loeb, and J. Gröger. 1998. Krill (*Euphausia superba*) density, proportional and absolute recruitment and biomass in the Elephant Island region (Antarctic Peninsula) during the period 1977–1997. *Polar Biology* 19: 393–398.

Siegel, V., B. Bergström, U. Mühlenhardt-Siegel, and M. Thomasson. 2002. Demography of krill in the Elephant Island area and its significance for stock recruitment. *Antarctic Science* 14: 162–170.

Sievers, H.A. and W.D. Nowlin. 1984. The stratification and water masses at Drake Passage. *Journal of Geophysical Research* 89: 10489–10514.

Smetacek, V., R. Scharek, and E.-M. Nöthig. 1990. Seasonal and regional variation in the pelagial and its relationship to the life history cycle of krill. In: *Antarctic ecosystems: ecological change and conservation* (eds., K.R. Kerry and G. Hempel), pp. 103–114.Springer, Berlin.

Smith, D.A., E.E. Hofmann, J.M. Klinck, and C.M. Lascara. 1999. Hydrography and circulation of the West Antarctic Peninsula continental shelf. *Deep-Sea Research I* 46: 925–949.

Smith, W.O., M.S. Dinniman, J.M Klinck, and E. Hofmann. 2003. Biogeochemical climatologies in the Ross Sea, Antarctica: seasonal patterns of nutrients and biomass. *Deep-Sea Research II Topical Studies in Oceanography* 50: 3083–3101.

Spiridonov, V.A. 1995. Spatial and temporal variability in reproductive timing of Antarctic krill *(Euphausia superba* Dana). *Polar Biology* 15: 161–174.

Spiridonov, V.A. 1996. A scenario of the Late-Pleistocene-Holocene changes in the distributional range of Antarctic krill (*Euphausia superba*). *Marine Ecology* 17 (1–3): 519–514.

Sprintall, J. 2003. Seasonal to interannual upper-ocean variability in the Drake Passage. *Journal of Marine Research* 61: 27–57.

Stein, M. 1986. Variability of water masses and current off the Antarctic Peninsula during SIBEX. *Arch Fischereiwiss* 37: 25–50.

Stein, M. 1988. Variation of geostrophic circulation off the Antarctic Peninsula and in the southwest Scotia Sea. In: *Antarctic ocean and resources variability* (ed., D. Sahrhage), pp. 88–91. Springer, Berlin.

Stein, M. 1989. Seasonal variation of water masses in Bransfield Strait and adjacent waters. *Arch Fischereiwiss enschaft* 39: 15–38.

Stein, M. 1992. Variability of local upwelling off the Antarctic Peninsula, 1986–1990. *Archiv Fischereiwissenschaft* 41: 131–158.

Turney, C.S.M. et al. 2004. Millennial and orbital variations of El Niño/Southern Oscillation and high-latitude climate in the last glacial period. *Nature* 428: 306–310.

Tynan, C.T. 1998. Ecological importance of the southern boundary of the Antarctic Circumpolar Current. *Nature* 392: 708–710.

Ward., P. et al. 2002. The southern Antarctic Circumpolar Current front: physical and biological coupling at South Georgia. *Deep-Sea Research I: Oceanographic Research Papers* 49: 2183–2202.

Ward, P., M. Whitehouse, M. Brandon, R. Shreeve, and R. Wood-Walker. 2003. Mesozooplankton community structure across the Antarctic Circumpolar Current to the north of South Georgia: Southern Ocean. *Marine Biology* 143: 121–130.

White, W.B., S.C. Chen, R.J. Allan, and R.C. Stone. 2002. Positive feedbacks between the Antarctic circumpolar wave and the global El Niño–Southern Oscillation. *Journal of Geophysical Research* 107: Art. No. 3165.

Witek, Z., W. Kittel, H. Czykieta, M.I. Zmijewska, and E. Presler. 1985. Macrozooplankton in the southern Drake Passage and in the Bransfield Strait during BIOMASS-SIBEX (December 1983–January 1984). *Polish Polar Research* 6: 95–115.

Yuan, X. and D.G. Martinson (2001). The Antarctic dipole and its predictability. *Geophysical Research Letter* 28: 3609–3612.

Zane, L., L. Ostellari, L. Maccatrozzo, L. Bargelloni, B. Battaglia, and T. Patarnello. 1998. Molecular evidence for genetic subdivision of Antarctic krill (*Euphausia superba* Dana) populations. *Proceedings of the Royal Society of London, Series B* 265: 2387–2391.

Zane, L. and T. Patarnello. 2000. Krill: a possible model for investigating the effects of ocean currents on the genetic structure of a pelagic invertebrate. *Canadian Journal of Fisheries and Aquatic Science* 57 (Supplement 3): 16–23.

Zwally, H.J., J.C. Comiso, C.L. Parkinson, D.J. Cavalieri, and P. Gloersen. 2002. Variability of Antarctic sea ice 1979–1998. *Journal of Geophysical Research* 107: Art. No. 3041.

Index

THE PETER YODZIS FUNDAMENTAL ECOLOGY SERIES

N. Rooney, K.S. McCann and D.L.G. Noakes (eds.): *From Energetics to Ecosystems: The Dynamics and Structure of Ecological Systems.* 2007 ISBN: 1-4020-5336-3

D.A. Vasseur and K.S. McCann (eds.): *The Impact of Environmental Variability on Ecological Systems.* 2007 ISBN: 978-1-4020-5850-9

MIX
Papier aus verantwortungsvollen Quellen
Paper from responsible sources
FSC® C105338

If you have any concerns about our products,
you can contact us on
ProductSafety@springernature.com

In case Publisher is established outside the EU,
the EU authorized representative is:
Springer Nature Customer Service Center GmbH
Europaplatz 3, 69115 Heidelberg, Germany

Printed by Libri Plureos GmbH
in Hamburg, Germany